Memoirs of a Proof Theorist

Gödel and other Logicians

Memoirs of a Proof Theorist

Gödel and other Logicians

Mariko Yasugi

Kyoto Sangyo University, Japan

Nicholas Passell

University of Wisconsin, Eau Claire, USA

World Scientific
New Jersey • London • Singapore • Hong Kong

Published by

World Scientific Publishing Co. Pte. Ltd.

5 Toh Tuck Link, Singapore 596224

USA office: Suite 202, 1060 Main Street, River Edge, NJ 07661

UK office: 57 Shelton Street, Covent Garden, London WC2H 9HE

British Library Cataloguing-in-Publication Data
A catalogue record for this book is available from the British Library.

MEMOIRS OF A PROOF THEORIST
Gödel and Other Logicians

ISBN 981-238-279-8

Printed by FuIsland Offset Printing (S) Pte Ltd, Singapore

Translators' Foreword

This book is a translation from Japanese of the September 10, 1998 revision of Gaisi Takeuti's <u>Gödel</u> first published in 1986 by Nippon Hyoron Sha,LTD, Tokyo, Japan. The author's prefaces explain the structure of the book. Each individual piece was written for the monthly Sugaku Seminar (Mathematics Seminar: published by Nippon Hyoron Sha,LTD) as an essay or a personal reflection.

Chapter 2 is originally from the Proceedings of the Sixth International Congress of Logic, Methodology and Philosophy of Science, Hannover 1979 ed. by L.J. Cohen, J. Łoś, H. Pfeiffer and K.P. Podewski, North-Holland, 1982. In Sugaku Seminar, it was translated and revised from the author's own manuscript by Akira Oide. We have reproduced the original here.

Appendix A was added in the 1998 edition. Appendix B is new and has been added at the author's request. It was published in Suri-Kagaku (Mathematical Sciences: published by Saiensu-sha Co.Ltd) in 1998, and has been tranlated here. It gives some history of second order proof theory together with a sense of the author's intellectual development as this theory was created.

Several chapters in <u>Gödel</u> have been omitted at the author's request, mainly because they were translations into Japanese from other languages.

Acknowledgements of authorization for reproduction are given at the end of this foreword. Sources of citations are given where they occur.

In most cases the translation has been kept close to the Japanese text, to maintain the flavor of the original, and to convey the sense of the time and the author's feelings, even at the cost of some awkwardness. In several cases this has resulted in the repetition of sentences and even entire paragraphs. Minor errors and inconsistencies were corrected where they were noticed.

Short footnotes have been added by the translators where it seemed they would facilitate the reader's understanding. Footnotes by the author are indicated as "Author's Note." Where more extensive clarification is necessary the reader is referred to the author's citations. The translators apologize for not supplying references for reasons of time and space.

The translators would like to thank: Professor Gaisi Takeuti for his patient assistance, his availability to answer questions and for a copy of his photograph; Tetsujiro Kamei and Nippon Hyoron Sha,LTD for a photograph of Kurt Gödel as well as encouragement of our doing this translation; the Translation Committee of the Association for Symbolic Logic especially the (then) Chair, Steffen Lempp, for suggesting and facilitating this translation; John N. Crossley for his advice on technical expressions; Susumu Hayashi for contributing mathematical and historical insight; and finally, Susan Johnson who has remained remarkably patient, cheerful and accurate through the many revisions of this manuscript.

The translators are indebted to the following copyright holders for granting permission to print the original texts and English translations.

Nippon Hyoron Sha,LTD, Tokyo, for most of the chapters from Gödel.

Elsevier Science, Amsterdam, for "Work of Paul Bernays and Kurt Gödel" by Gaisi Takeuti, pp.77-85, in Proceedings of the Sixth International Congress of Logic, Methodology and Philosophy of Science, Hannover 1979, ed. by L.J. Cohen, J. Łoś, H. Pfeiffer and K.P. Podewski, published by North-Holland (1982).

Cambridge University Press, Cambridge, for "Supplement to the second edition [1963]" by Kurt Gödel, pp.482-485, in Philosophy of Mathematics, 2nd ed., ed. by P. Benacerraf and H. Putnam (1984).

Springer-Verlag, Heidelberg, for "Tribute to Dr. Gödel" by John von Neumann, pp.IX-X, in Foundations of Mathematics, ed. by J.J. Bulloff, C. Holyoke and S.W. Hahn (1969).

Saiensu-sha Co.,Ltd, Tokyo, for "Birth of Second Order Proof Theory by the Fundamental Conjecture on GLC" by Gaisi Takeuti, pp.10-17, in Extra Issue of Suri-Kagaku, 'Several Topics on the History of Modern Mathematics' (1998).

There are many short citations, whose sources are clearly stated where they occur, either by the author or translators.

September 30, 2002
Mariko Yasugi
Nicholas Passell

Preface to the Original Edition

It has been arranged that what I have written about Gödel from time to time is to be printed in a single book. I feel that I have written everything that I have to say about Gödel. In that sense, what is printed here expresses my whole image of Gödel.

Most of what has been collected here was published in the magazine "Sugaku Seminar"[1]. The dates of publication are written at the end.[2] I have made minor corrections to the original manuscripts, but have not made any change to the content itself. It is because I wished to respect my feelings of the time of first publication. I have also left the items about logicians other than Gödel whom I wrote about along with Gödel. I believe that I can convey some aspect of Gödel indirectly from them by preserving the original style rather than abandoning them. As time has elapsed, there have been certain changes. For example many of the participants have died since. However, that has not been mentioned in the text. Instead, I have appended a chronological table at the end of this volume.

I would like to express my heartfelt gratitude to Mr. Tetsujiro Kamei of Nippon Hyoron Sha,LTD, whom I owe a great deal in publishing this book.

June 24, 1986
Gaisi Takeuti

---

[1]Mathematics Seminar: published by Nippon Hyoron Sha,LTD, Tokyo, Japan
[2]before the main text in this volume

Preface to the Revised Edition

In publishing a revised edition, I have added several manuscripts which I have written after the first edition. They are 'Set Theory and Related topics,' 'From Hilbert to Gödel,' 'Axioms of Arithmetic and Consistency,' 'A Report from Gödel '96,' and 'Having Read "Gödel Remembered"'. All of them are either directly related to Gödel or to his works.

Further I have added as an appendix 'On Gödel's Continuum Hypothesis'. this is close to a mathematical article and so may not fit the style of this book. I nevertheless decided to add it, as it has direct relationship to the problems which appear at various places in the book.

Each chapter reflects the time of its writing. The interested reader may refer to the chronology list of publications at the end.[3]

During the preparation of the revised edition, Mr. Shin Yokoyama of Nippon Hyoron Sha,LTD carefully examined the manuscript, from which I benefited a great deal. Let me express my gratitude towards him.

June 5, 1998
Gaisi Takeuti

---

[3]before the main text in this volume

Preface to the Revised Edition

Author's Preface to the Translation

I am pleased that my book "Gödel" has been translated and is to be published by World Scientific.

I have briefly described the structure of this book and how I approached the material in the prefaces of the first and the revised editions in Japanese.

I wrote of my experiences with Gödel and other logicians as forthrightly and accurately as I could in order to convey the feelings and events in the world of logicians in our era. I hope it will be of some interest to mathematicians and logicians of the 21st century.

I am grateful to the translators for having taken their time. They understood my feelings towards the articles and tackled the difficult work of preserving my feelings and my style of writing, while at the same time making the prose readable. They also examined the content carefully and pointed out some insufficiencies and errors, which I corrected (in Japanese).

The second Appendix has been added to this translation. In the main text, I mentioned "My Fundamental Conjecture" without explanation. I thought it might be of help to young logicians if some explanation of it were added.

September 30, 2002
Gaisi Takeuti

Curriculum Vitae of Gaisi Takeuti

1926: Born in Ishikawa Prefecture, Japan

1947: Graduated from Department of Mathematics, Faculty of Science, University of Tokyo, Tokyo, Japan

1962: Professor, Tokyo Kyoiku University, Tokyo, Japan,

1966-1996: Professor, University of Illinois, Urbana-Champaign, Illinois, USA

1981: Awarded Asahi Prize, Asahi Newspaper Company, Tokyo, Japan

Major interests in research: Proof theory, Boolean valued analysis, bounded arithmetic

Gaisi Takeuti is a world renowned proof theorist, who has made pioneering contributions in the area of consistency proofs of systems of second order arithmetic. He has written numerous research articles and books.

Gaisi Takeuti

Kurt Gödel

Contents

Original Publication Dates

All the articles except the one which became Chapter 2 were originally written in Japanese. Chapter 2 reproduces an article written in English.

All the chapters except Appendix B were included in the original book "Gödel" by Gaisi Takeuti (revised edition), whose translation this book is, as a collection of essays and articles which had been published elsewhere (except Appendix A, which was printed in "Gödel" for the first time). A list of the original publication dates is given below.

"S.S." abbreviates "Sugaku(Mathematics) Seminar", a magazine published by Nippon Hyoron Sha,LTD, Tokyo, Japan; "S.K." abbreviates "Suri-Kagaku" (Mathematical Sciences), a magazine published by Saiensu-sha Co.Ltd, Tokyo, Japan.

Title	Source	Month/Year
On Gödel	S.S.	December 1985
Work of Paul Bernays and Kurt Gödel	S.S.	February 1980
Hilbert and Gödel	S.S.	March 1980
Short Biographies of Logicians	S.S.	April ~ June 1969
Set Theory and Related Topics	S.S.	January 1989
From Hilbert to Gödel	S.S.	March 1992
Axioms of Arithmetic and Consistency	S.S.	February 1994
A Report from Gödel'96	S.S.	January 1997
Having read "Gödel Remembered"	S.S.	February 1988
A Tribute to the Memory of Professor Gödel	S.S.	May 1978
On Gödel's Continuum Hypothesis	Gödel	September 1998
The Birth of Second Order Proof Theory...	S.K.	December 1998

Note. 1) "Work of Paul Bernays and Kurt Gödel" was first published in Sugaku Seminar in 1980, translated from the author's manuscript by A. Oide. In the 1998 edition of "Gödel" a translation by A. Oide of the article below was contained: Proceedings of the Sixth International Congress of Logic, Methodology and Philosophy of Science, Hannover 1979 ed. by L.J. Cohen, J. Łoś, H. Pfeiffer and K.P. Podewski, North-Holland, 1982. Here we have reproduced the 1982 version. We have, however, preserved the sections of the 1998 edition.

2) "From Hilbert to Gödel" was originally entitled: "Hilbert and Gödel."

Chapter 1

On Gödel

I was recently asked by Kreisel to write about Gödel, but have just declined the request. (Kreisel is a distinguished logician, who was closest to Gödel during the 1960s.)

Paraphrasing Kreisel's request, various people who did not know Gödel well enough have written about him in recent years. The person they portray is entirely different from the true Gödel. Those, like you and me, who have been close to Gödel are obliged to convey his true personality. I will certainly write myself, but what you have to say must be different from what I do. So I would like you to write as well.

The reason I have refused this offer is that I do not have time. I am honest. If I tried to comply with Kreisel's request, I would have to check various facts before writing, even relating to circumstances I am familiar with. At present, I have no time for it.

There are, however, other reasons as well. I reported already on Gödel's mathematics at the International Conference on Foundations of Science in 1979. (See "Work of Bernays and Gödel" in this volume.) Otherwise, there are only personal recollections remaining. Another reason is the sadness of Gödel's last days.

The last occasion on which I spoke with Gödel was a long distance call from London. While I was in England attending a conference, Tennenbaum called and, failing to reach me, left a message saying "Please give a call to Gödel at any cost!"

The person who answered the telephone was not the Gödel I had known. Instead, there was a Gödel whose only wish was that death come as soon as possible without causing any trouble to others. There was no grief, no sorrow. What consumed him was such a nihilistic despair that it could scarcely be called despair.

1

After his death, his last days lingered in my mind. Why should the last days of Gödel have been like this? Furthermore, when his wife returned home from his funeral, a burglar had broken in and jewelry and other goods had been stolen. I felt then resentment against modern times as well as American society.

1.1 My fundamental conjecture and Gödel

Except for the last two or three years, the Gödel I knew was a warm and tender person.

It was in 1959 that I met Gödel for the first time, while I was staying at the Institute for Advanced Study in Princeton. Although there was a rumor at the time that Gödel was awkward with people and was isolated, I heard that a year before my visit to the States, there had been several logicians at the Institute, Kleene among them, and Gödel had held a regular logic seminar weekly. The Gödel I met for the first time was a cheerful person.

Gödel became deeply interested in the problem I was concerned with at that time — my Fundamental Conjecture[1]—.

This typified Gödel's attitude towards mathematics throughout his life. He would show strong interest in any significant problem which was proposed regardless of origin, would review its inner meaning from his standpoint, and then would present new ideas of his own concerning the problem.

His Completeness Theorem was a solution to a fundamental problem which had been proposed in the textbook by Hilbert and Ackermann, and his Incompleteness Theorem was his quiet answer to Hilbert's Program, especially to the metamathematical elements of the program, whose goal was to develop a new arithmetic.

Furthermore, it might be claimed that his work in set theory questioned the meaning of Russell-Whitehead's logicism, and salvaged the essence of the continuum hypothesis through the study of Hilbert's Program on it.

Gödel also studied Gentzen's works thoroughly and pondered them.

His view on my fundamental conjecture was that counter-examples would be discovered by the method of his Incompleteness Theorem or by some nonstandard methods. As my English was rather poor, he would immediately take me to the library when I asked questions, take out periodicals and indicate the articles and specific points in question. At that

---

[1]Takeuti's fundamental conjecture sates that, for the second order or the higher order predicate calculus, a theorem can be proved without detour: See Appendix B.

time, I was concentrating on my own problem, and I did not have comprehensive knowledge of logic. Thus, in the beginning, Gödel shared his knowledge with me.

His way of teaching nonstandard models was an interesting one. It went as follows. Let T be a theory with a nonstandard model. By virtue of his Incompleteness Theorem, the consistency proof of T cannot be carried out within T. Consequently, T and the proposition "T is inconsistent" is consistent. There is, therefore, a natural number N which is the Gödel number of a proof leading to a contradiction from T. Such a number is obviously an infinite natural number.

It was also he who taught me "large cardinals." His way of teaching was a very thorough one. For example, having conveyed a broad view, he would tell me further that I should visit Bernays and ask him about various matters.

Now, returning to my fundamental conjecture, Gödel appeared to think that, if its proof were to be extremely impredicative, then there would be a counter-example and the state of things similar to the Incompleteness Theorem would hold. In fact, what is interesting about my fundamental conjecture is that, if one admitted an extremely impredicative demonstration, then it would become trivial. When I told this to Gödel, he was very surprised and he seemed to become more interested in the problem. Gödel asked me if I had published that fact, and I replied that I had not because it was trivial. He then said: "Publish it by all means! Many people will change their views on your conjecture." As a result, I published a trivial article against my will.

It seemed that Gödel became sincerely interested in my program of consistency proofs through the fundamental conjecture. I was moved when I found later numerous little comments and questions in the reprints of my articles on the fundamental conjecture as well as on "ordinal diagrams", a tool for the proof of my conjecture, which apparently Gödel had written in as he had studied them. The reader may not believe it, but I think that hardly anybody in the world except my students, Schütte and his disciples read my articles on this subject seriously. Notably, Schütte's interest in my work was the result of Gödel's influence, as I shall explain.

Gödel thought that, for the progress of my fundamental conjecture, it would be useful to put Schütte and myself together, and so he invited Schütte to the Institute for Advanced Study. When I dropped by at the Institute one late summer day, a stranger came directly to me, and asked "Would you know Takeuti?" I replied that I was the person himself and then he introduced himself as Schütte and told me: "I have just talked with Gödel, and found that Gödel is interested in your fundamental conjecture" and so on and so forth. I imagine Gödel told him that there might be some kind of relationship between what Schütte had worked on and my fundamental conjecture, and he suggested some research directions to him.

Schütte was a person other than Gödel who became interested in my work. I think Schütte's interest was aroused by Gödel. Looking back, we can say that Gödel's foresight was correct, considering that the results on my fundamental conjecture by Motoo Takahashi and Prawitz were based on Schütte's work done at the Institute at that time.

Thanks to Gödel, during the two years of my stay at the Institute, many logicians such as Bernays, Schütte, and Feferman were there. Smullyan and Putnam were at the University as well. They held a logic seminar every week, and the logic group was very lively. In particular, there were two proof-theorists in the rare Gentzen style together (Schütte and Takeuti), and so we were high-spirited. Smullyan would make me laugh by referring to us in a joking manner: "Is your name TakéSchütte?" It is my sense that the energy and morale of logicians in Princeton was due to Gödel's kindness, cheerfulness and warmth.

At the end of my two years' stay at the Institute, Gödel suggested that I visit Tarski in Berkeley. I felt sad to leave Princeton and Gödel, as I had been so happy there, and I made my way to Berkeley reluctantly. Considering it now, Gödel must have thought it would be important for me to step out into a world different from his.

Subsequently, I returned to Japan once and then came back to America. Gödel's presence was an important factor in the motivation for this. Because of him, I visited the Institute twice after my return to America. During my second stay, I spoke with Gödel quite often. I was rarely in my office, and so Gödel would phone me at home and ask me "When in the world are you in your office?" It seemed that he felt awkward if somebody other than I answered him when he phoned, as he was a shy person. Nevertheless, when my youngest child, who was a grade school girl, answered, he was in good humor and left a message.

Of discussions during this period, there is surprisingly little left which I can recollect in detail. One reason is perhaps that I immediately assimilated the content of our discussions and so they became part of me. More accurately expressed, however, it was because until my first visit I had been confined in the solitary and isolated world of my fundamental conjecture, while Gödel had extensive knowledge. Such a circumstance created a dramatic state of affairs, so to speak, where two almost disjoint worlds met. On the other hand, as my understanding grew on my second and third visits, we were able to discuss various problems on common ground with much shared knowledge.

I certainly knew Gödel for a long time, and so I have many memories concerning him, but not of any specific incidents, and so I do not think I should write about them.

1.2 Gödel's mathematics

For almost everyone in the mathematical community Gödel's discoveries must have seemed a miraculous flowering of genius.

Gödel first answered an open question in the foundations of mathematics, the Completeness Theorem, proposed by Hilbert and Ackermann in their text. Gödel resolved this problem and immediately thereafter proved his great Incompleteness Theorem, which at the time was a shocking result. Even today its significance is still being explored.

It took several years for the next important results to emerge. Gödel began working on what became "The Consistency of the Axiom of Choice and the Continuum Hypothesis" soon after finishing the Incompleteness Theorem but it took a great deal of work to reach this next group of remarkable results.

What is the secret of this Gödel's mathematical achievements?

To start, let us consider the Incompleteness Theorem. Prior to Gödel the mainstream approach to the Foundations of Mathematics consisted of two parts, that is, to regard mathematics as a set of rules on the arrangement of symbols and to develop a new arithmetic associated to it. Taken together, this might be called Hilbert's Program. In some respects we might describe Gödel's work as having carried out Hilbert's Program more thoroughly than any preceding attempt. How then is it the case that the brightest lights of the mathematical world who clustered around Hilbert were unable to do what Gödel did?

The answer is that, having committed themselves to the ideological slogan that mathematics is a set of rules governing arrangements of meaningless symbols, they rendered themselves incapable of taking advantage of the content of the mathematical substance on which they worked. Gödel on the other hand was able to reach the Incompleteness Theorem by applying metamathematical insights in parallel with the substantial content of the mathematical systems of set theory, classical number theory, etc., with which he worked.

This was an astounding surprise to the Hilbert school. Before Hilbert the central obsession of those who worked in the area of foundations of mathematics was the meaning of symbolism. The Hilbert school took the new viewpoint of regarding mathematics as arrangements of meaningless symbols and they developed its associated metamathematics as a new arithmetic. This heralded the arrival of a new era.

Gödel showed as a consequence the impossibility of carrying out Hilbert's Program in the naive way that Hilbert had intended. His success unified the two viewpoints.

For those mathematicians who were committed to Hilbert's approach, Gödel's result achieved the exact opposite of their objective using their own method, fully developing it and using it ingeniously in a manner directly opposed to their ideology. This was a double blow.

Nonetheless, it was the Hilbert school who best understood Gödel's work. Indeed von Neumann, then a bright young man belonging to Hilbert's inner circle, admired Gödel's work and made an effort to bring him to the world's attention. As an example, in the paper on the Incompleteness Theorem, Gödel presented only a short outline of the second theorem demonstrating that the consistency proof could not be carried out within the system in question, with a promise of a detailed proof in a future paper. This was actually carried out first in the second volume of Hilbert and Bernays' text "Grundlagen der Mathematik." Additionally, many further extensions of Gödel's work began from this book.

This dual approach taken by Gödel in his proof of the Incompleteness Theorem is also seen in his epoch making work establishing the consistency of the axiom of choice and the continuum hypothesis. Gödel obviously started from Russell and Whitehead's logicism as well as Hilbert's Program on the continuum hypothesis.

Russell and Whitehead claimed that a contradiction in set theory arose from the definitions of non-elementary sets, which they called "impredicative." They believed that this difficulty could be avoided if the construction of new sets were confined to elementary processes. However when the idea was put into practice it failed to produce the desired range of mathematical structures and theorems.

Hilbert, on the other hand, proposed to resolve the continuum hypothesis by obtaining all the subsets of the natural numbers with constructive and transfinite repetitions of elementary definitions of sets. His article has an optimistic tone suggesting that by executing the program he has out-lined he should succeed soon, if not immediately. Thus we see that Russell, Whitehead and Hilbert had an ideological commitment to construction at the core of their programs which was the source of both their programs' strengths and their ultimate limitations.

Gödel achieved his breakthroughs by carrying out the constructive ideas of Russell, Whitehead and Hilbert within non-constructive set the-ory. Specifically Gödel executed Russell-Whitehead's idea and Hilbert's more advanced one of constructions within transfinite set theory. Again, Gödel's dual approach is clearly seen. Russell-Whitehead restricted set def-initions to elementary ones because they saw no philosophical justification for transfinite set theory. If they could have accepted transfinite set theory, their standpoint would have been irrelevant. Hilbert's approach of repeti-tion of elementary constructive and transfinite definitions of sets was taken because it made the cardinality of the sets thus constructed clear. In both approaches then, the constructive and elementary viewpoint was believed to have great philosophical significance.

Gödel questioned the program of Russell-Whitehead philosophically and applied their constructive ideas and those of Hilbert to transfinite set theory, its philosophical opposite. Thus Gödel alone succeeded spectacularly by merging two standpoints, strict constructivity and transfinite methods, that is, using the constructive approach in a transfinite manner.

What is this dual approach of Gödel's? There is no secret here. Gödel pondered the fundamental problems and recognized that the mathematical methods which were apparently attached to ideological programs could be separated from their ideological baggage. Then he made use of these meth-ods in the best possible way. One way of saying it is that Gödel transcended his times. It took twenty years for the field to catch up with him, though the war also contributed to the delay.

This delay seems to have influenced Gödel's mathematics. Gödel be-came isolated. One factor may have been his shyness, though it is known that he was on friendly terms with Einstein and von Neumann. Had the world of mathematical logic reached Gödel's level sooner, his communi-cations with other logicians would have been more productive. As it was, Gödel associated most closely with Einstein and Von Neumann, whose prin-cipal areas of interest were not logic or set theory.

One consequence was that Gödel stopped publishing his work. For example, applications of his axiom of constructibility to descriptive set the-ory included some remarkable results. Nonetheless, he only communicated them individually. There are other examples as well of investigations carried out with enthusiasm which went unpublished, e.g. properties of various sets and hypotheses concerning the continuum hypothesis and interpretations of some new logical symbols. All these were written in his notes in short-hand and never announced to the public. If he had published this material without reservation, others would have studied it and that in turn would have induced him to do more great work. I regret things went otherwise.

As can be seen from the first three outstanding results, even Gödel was influenced by the ideas of others. Isolation, no matter how honorable, does not encourage productivity even from as deep a mind as Gödel's.

By the time Gödel's colleagues finally had assimilated his work and significant new material began to be published which could stimulate Gödel, he was almost sixty years old. Of course this was a consequence of his having been so far ahead of his contemporaries, but it was a sad fact. Following his early epoch-making achievements, Gödel did do a considerable amount of interesting research including work on the consistency of number theory and in general relativity. He also left much unpublished work. None of this is comparable however with the first three miraculous results. That was the misfortune of a person who was too far ahead of his time.

1.3   The Gödelian "Boom"

As far as I know, there have been three occasions when Gödel gained broad attention. The first occurred when a philosopher published a book called "The Proof of Gödel!" The excitement did not reach Japan but it was fairly widespread. The second was the result of IBM publicity including photos, which did reach Japan. The third came with the publication of Hofstadter's "Gödel, Escher, Bach."

"Gödel, Escher, Bach" seems to have been the most flamboyant pro-moter of attention to Gödel. It seems to have connected to the current fascination with computers and information technology. I do not know what sort of impact this attention will have. But it is a good thing that there is a chance for Gödel's work to be understood. I hope this understanding is genuine. I do not have any idea what connection there is between Gödel and Escher or Gödel and Bach.

Recently a student of mine who had just studied Gentzen and Gödel told me, "Gentzen is Bach, and Gödel is Debussy." This must have expressed the impression the student received from the clarity of Gentzen's writing on the one hand and Gödel's ideas as well as his style on the other. Everyone seems to relate to Gödel subjectively.

There are many composers and artists whom I like, but I don't perceive an affinity between Gödel and any of them. Gödel is Gödel through and through and not akin to anyone else.

Chapter 2

Work of Paul Bernays and Kurt Gödel

In the last two years, we lost two great logicians Paul Bernays and Kurt Gödel. Both Bernays and Gödel worked in proof theory and set theory and were deeply concerned over the foundation of mathematics. [1]

2.1 Bernays as a collaborator of Hilbert

Paul Bernays' logic career started in 1917, when Hilbert invited Bernays to Göttingen to work with him on the foundation of mathematics. From then till 1933, Bernays was Hilbert's coworker in Göttingen, the center of mathematics at the time. In 1904, Hilbert had proposed Hilbert's Program to save mathematics from the crisis caused by contradictions in set theory. Hilbert's Program consists of two parts.

1) *Formalization of mathematics.*

This is to make a mathematical theory a formal system in the following way.

First enumerate all primitive symbols in the theory. Then describe what kind of combination of symbols is a meaningful statement in the theory. Such kind of combination of symbols is called a formula. Finally describe what kind of sequence of formulas is a proof of the theory. A proof thus formalized is a concrete figure of symbols and is called a proof-figure.

2) *Consistency-proof of a formal theory.*

This is to provide a foundation for mathematics for proving its consistency. Here the meaning of consistency is the consistency of a formal theory, i.e. nonexistence of a proof-figure ending in a contradiction $0 = 1$. Since a proof-figure is a concrete figure of symbols, Hilbert thought that his

---

[1] Here we reproduce Takeuti's original article faithfully except for division into sections and some typos. See 1) of Note in "Original Publication Dates."

finite standpoint, i.e. a standpoint using only intuitive combinatorial arguments on concrete figure, is sufficient for the consistency proof. Hibert's finite standpoint is more precisely the following : We can finitely operate on a concrete figure given before us, and infer a general statement as a Gedanken experiment.

The first part of Hilbert's Program, i.e. the formalization of mathematics, had been already done at the time, e.g. in Whitehead and Russell's *Principia mathematica*. Therefore the second part is the actual Hilbert's Program. For this purpose, one has to develop metamathematical theory of proof-figures based on the finite standpoint. Such a theory is called *proof-theory* or *Beweistheorie* in German. The joint work of Hilbert and Bernays of course belongs to proof-theory. Their work was written in detail in their two books *Grundlagen der Mathematik* (Springer, I, 1934 and II, 1939). The books were actually written line by line by Bernays. The main body of the books concerns the finite standpoint, the basic properties of the first order predicate calculus, ε-calculus, a proof-theoretic version of Gödel's completeness theorem, and the detailed proof of Gödel's incompleteness theorem. I will discuss Gödel's theorems later but it should be remarked here that Gödel only outlined his proof of the second part of his incompleteness theorem and its detailed proof was first carried out in Hilbert-Bernays' book.

First of all, I would like to say that the first chapter of volume 1 written by Bernays was the first coherent statement of what one might call Hilbert's Program. As the most original part of the books, I would like to explain their theory of ε-calculus. In ε-calculus, we introduce ε-symbol in addition to the usual language of the first order predicate calculus. Then we add the following formation rule for terms. If $A(a)$ is a formula with a free variable a, then $\varepsilon x A(x)$ is a term. The intended meaning of $\varepsilon x A(x)$ is some x which satisfies $A(x)$ if there exists a such. This can be expressed formally only by the following axiom schema:

$$A(t) \rightarrow A(\varepsilon x A(x))$$

where t is an arbitrary term. If one uses ε-symbol, he can eliminate quantifiers by replacing $\exists x A(x)$ by $A(\varepsilon x A(x))$. On the other hand, if a theorem without ε-symbols is provable in ε-calculus, then a method is given to transform a proof-figure of the theorem in ε-calculus into a proof-figure without ε-symbol. They developed proof-theory on ε-calculus and proved Herbrand's theorem as an application of their theory. Today ε-calculus is scarcely used. For example, I use Gentzen's sequential calculus in the place of ε-calculus. However, by looking at their books, I am convinced that Hilbert and Bernays developed the ε-method to such a degree that one can solve a problem on the first order predicate calculus by their theory as easily as by any other methods. In other words, they essentially completed the proof-theory of the first order predicate calculus.

2.2 Set theory and proof theory of Bernays

In 1934, the political situation made Bernays go back to Zürich, where he taught 25 years in E.T.H. (The Eidgenössische Technische Hochschule). From 1937 to 1954 Bernays published seven papers on axiomatic set theory in the Journal of Symbolic Logic. A characteristic of this theory is that it has the class variables in addition to the set variables. As for his work on set theory, it is fair to say that he is the first person who organized set theory in the present standard and presented it in such a way that people could learn and use it very easily. For example, Gödel used Bernays' presentation of set theory in his famous monograph and many of us learned set theory from Gödel's monograph. In addition to the elegant presentation of set theory, Bernays showed which parts of axioms in set theory are used for what part of mathematics. For example, he introduced the axiom of dependent choice as the part of the axiom of choice necessary to analysis.

Evidently, a wide range of logicians profited from his work on set theory. In proof-theory, Bernays' influence is even greater. Kreisel's no-counter-example-interpretation is a continuation of the Hilbert-Bernays' ε-theorem. One cannot think of Feferman's work on arithmetization of metamathematics without Hilbert-Bernays' work of Gödel's theorem. Though Gentzen's teacher was Hermann Weyl, Gentzen's letters show that Bernays is his teacher in the true sense of being the one who encouraged him, advised him, and listened to him. Schütte, Prawitz, Maehara and myself among many others have worked in Gentzen's line. I would like to add one more important remark on Bernays' work. He made a great contribution to the philosophy of mathematics. Let me try to explain his philosophy of mathematics a little bit. He wrote "Mathematics, however, can be regarded as the theoretical phenomenology of structures." (Comments on *Ludwig Wittgenstein's Remarks on the Foundations of Mathematics*, Ratio II, No. I (1959), pp. 1-22). In his opinion, the main role of set theory is to provide us with models of structures. It seems to me that the superiority of Bernays' philosophy consists in avoiding the kind of simple-mindedness

common among others. Bernays was the first to insist on the deceptive-
ness of the so-called conflicts between Brouwer's and Zermelo's or Cantor's
views: when Brouwer spoke of contradictions with classical mathematics,
Bernays pointed out that the logical notions and the ranges of variables
(choice sequences) had different meanings. So the 'conflict' consisted in
different views of what is worth studying. Bernays also was the first to
insist that Hilbert's finitistic mathematics was a restriction of intuitionistic
mathematics. His aphorism was "classical mathematics is the mathematics of Being, intuitionistic mathematics is the mathematics of Processes."
Dealing with the criticism of classical mathematics advanced by the intu-
itionists, he wrote "Heyting argues that it is too naive to ask questions
like 'Does there really exist a well-ordering of the continuum ?' But the
question need not be put in this way. The question rather may be asked:
'Is it suitable to adopt the strong extrapolation of classical analysis as it
is formulated in the conceptions of set theory, in particular in the theory
of transfinite cardinals ?' At the present time no deciding experience has
been found to answer this question. Mathematicians have different opinions
about the suitability of stronger or only weaker methods of idealizations.
At all events the fact that our mathematical idealizations are so successful
is a striking mental experience and by no means something trivial." (*Mathematics as a domain of theoretical science and of mental experience*, Logic
Colloquium '73, pp. 1-4, Edited by H. E. Rose and J. C. Shepherdson,
North Holland, 1975.) Bernays' philosophy is much more subtle than the
standard literature.

2.3   The Completeness Theorem of Gödel

In 1930, Gödel proved the completeness theorem in his Ph.D. thesis which
was later published as *Die Vollständigkeit der Axiome des logischen Funk-
tionenkalkuls*, Monatshefte für Mathematik und Physik, vol. 37, pp. 349-
360, 1930. Previously in 1928, Hilbert and Ackermann formulated the first
order predicate calculus in their book *Grundzüge der Theoretischen Logik*
and stated : It is still an unsolved problem if the axiom system is at least
complete in the sense that all logical formulas which hold in every structure,
can be deduced. It can only be said empirically that this axiom system has
sufficed in every application.

   Gödel proved this affirmatively, i.e. that the first order predicate calculus is indeed complete. There is an interesting fact about this theorem

as a historical event. In the eye of today's logician, the completeness theorem is an immediate consequence of a lemma of Skolem in his 1922 paper: *Einige Bemerkungen zur axiomatischen Begründung der Mengenlehre*, Wissenschaftliche Vorträge gehalten auf dem Fünften Kongress der Skandinavischen Mathematiker in Helsingfors vom 4. bis 7. Juli 1922, Helsingfors, 1923, pp. 217-232. But Skolem did not prove the completeness theorem and Gödel did. The completeness theorem can be expressed in the following way : If a sentence is consistent, then it has a model. What Gödel proved was the following stronger form: If countably many sentences are consistent, then they have a countable model.

2.4   The Incompleteness Theorem of Gödel

In 1931, Gödel proved the incompleteness theorem in his paper: *Über formal unentscheidbare Sätze der Principia Mathematica und verwandter Systeme*, I, Monatshefte für Mathematik und Physik, vol.38, pp.173-198. There he considered the formal system of Principia Mathematica and proved:

(1) If the system is ω-consistent, i.e.   for no $A(x)$ are all of $A(0), A(1), A(2), \cdots$ and $\neg \forall x A(x)$ provable in the system, then there exists an arith- metical sentence ϕ, such that either ϕ or $\neg \phi$ is not provable[2] in the system. (Rosser later improved the result by replacing ω-consistency by consistency.)

(2) If the system is consistent, then there exists an arithmetical statement which expresses the consistency of the system and is not provable in the system.

Here one has to talk about the so-called *arithmetization of metamathematics* in order to explain how one finds an arithmetical statement expressing the consistency of the system. In the arithmetization of metamathematics, we first represent primitive symbols by natural numbers. Then formulas and proofs are represented by sequences of natural numbers and sequences of sequences of natural numbers. Since we can enumerate all the finite sequences of natural numbers, we can represent formulas and proofs by natural numbers called their Gödel numbers in such a way that different formulas or proofs correspond to different numbers. What Gödel did was to carry out an analysis showing that metamathematical properties like 'x is a formula', 'x is a proof', 'x is a provable formula', etc. can be expressed by arithmetical formulas about the Gödel number x in the above corre-

---

[2]This should read: neither ϕ nor $\neg \phi$ is provable.

spondence. It follows that there exists an arithmetical statement which expresses the consistency of the system. It is because the system includes number theory that one can do many metamathematical deduction like mathematical induction on formalized metamathematical statements inside the system. Gödel obtained the above analysis in this way. By using it together with a diagonal argument, he succeeded in obtaining his results.

2.5   Gödel's method

We can see here a typical Gödel work. Gödel used a formal system in two different ways. On one hand, he used it as merely rules on sequences of meaningless symbols. On the other hand, he used its mathematical meaning to carry metamathematics inside it.

After a little while, it was clear that his method went through on any natural axiomatic system including Peano's arithmetic.

Even today, Gödel's result is extremely impressive. It certainly gives us a deep insight on the nature of axiomatic system and consequently on our knowledge itself.

But it was even more sensational in the following reasons.

> (1) Hilbert together with many others took it for granted that there exists an axiomatic theory formalizing mathematics in such a way that for every arithmetical statement either the statement or its negation is provable in the system. Gödel's result showed that this belief is completely false.
>
> (2) Since the finite standpoint Hilbert originally conceived was rather elementary, it is obvious that any arguments from his finite standpoint can be formalized in Peano's arithmetic and therefore Hilbert's Program, the central problem at the time, cannot be carried out from the finite standpoint Hilbert originally had in mind.

However, Gödel wrote at the end of his paper that his result does not rule out the possibility of finding some finitary consistency proof for mathematics in Hilbert's standpoint. Hilbert's standpoint merely requires the existence of a finitistic consistency-proof which cannot be represented in P (or in set theory or classical mathematics).

I do not know what Gödel meant by this remark. But I would like to add several comments here. In the finite standpoint—taken in a sense richer than Hilbert'sε—we can think of a concrete infinite sequence of concrete figures given before us instead of a single concrete figure and operate finitely on finite initial part of the sequence. Furthermore we can finitely operate on concrete operations, concrete operations on concrete operations, etc. and infer a general statement about them, as a *Gedanken experiment*. Actually in his 1936 paper: *Die Widerspruchsfreiheit der reinen Zahlentheorie*, Mathematische Annalen, vol.112, pp.493-565, Gentzen proved the consistency of Peano's arithmetic by using the accessibility up to the first ε-number and his proof of the accessibility can be justified in the finite standpoint described above.

Nevertheless Gödel's result made the original Hilbert's Program cease to exist. However, two more remarks should be added here.

> (1) Gödel's notes left in the Institute for Advanced Study show that Gödel studied Gentzen's work very carefully. Actually, in his 1958 paper: *Über eine bisher noch nicht benutzte Erweiterung des finiten Standpunktes*, Dialectica, vol.12, no.47/48, pp.280-287, Gödel himself published a consistency-proof which is different from Gentzen's proof but deeply related to it.
>
> (2) From 1959 till 1974, I had many occasions to converse with Gödel. What he wished to discuss with me was mainly consistency-proof. I had an impression that he was seriously interested in the subject. Also he was the logician who encouraged my consistency work most.

2.6 Gödel's set theory

In 1938, Gödel proved the consistency of the axiom of choice and the generalized continuum hypothesis with the axiom of set theory. (*The consistency of axiom of choice and the generalized continuum-hypothesis*, Proceeding of the National Academy of Science, vol.24, pp.556-557, 1938. *Consistency-proof for the generalized continuum hypothesis*, Ibid, vol.25, pp.220-224, 1939. *The consistency of the axiom of choice and of the generalized continuum-hypothesis with the axioms of set theory*, Annals of Mathematics Studies, no.3, Princeton University Press, Princeton, NJ 1940.)

The result is very important. However, as Gödel himself believed (cf. S.C. Kleene: *An Addendum to "The Work of Kurt Gödel"*, Journal of

Symbolic Logic, vol. 43, 1978, p. 613), his main achievement is his introduction of constructible sets. Gödel's notion of constructible sets comes from Whitehead and Russell's *ramified hierarchy*. Whitehead and Russell analyzed contradictions in set theory and found the vicious circle in the concept of sets as the source of contradictions. Thus they were lead to a concept of sets which is free from the vicious circle and very well justified philosophically. Their concept was called the ramified hierarchy. To explain it, let D be a well-defined set of well-defined objects like the set of all natural numbers. Let $x, y, \cdots$ range over the element of D. Since D is well defined, the notion $\forall x$ and $\exists y$ are also well-defined. Consequently, if $\phi(x)$ is a formula of the first order predicate calculus starting with well-defined predicates on D like $<$ for natural numbers, then a set of the form $\{x|\phi(x)\}$ is well-defined. Call these sets the *sets of ramified type* 1 and introduce variables $X^1, Y^1, \cdots$ ranging over the set of ramified type 1. Now new quantifiers $\forall X^1$ and $\exists Y^1$ are also well-defined. Therefore the set of the form $\{x|\phi(x)\}$ is well-defined if $\phi(x)$ is obtained by adding $\forall X^1$ and $\exists Y^1$ to the first order language. Call these sets the *sets of ramified type* 2, and so on for every finite type n. Whithead and Russell tried to construct classical analysis by using the ramified hierarchy. They could not do so and introduce the axiom of reducibility stating that any set of higher ramified type is already among the sets of the first ramified type. But the introduction of the axiom of reducibility changes the meaning of the ramified hierarchy and the original idea of the ramified hierarchy failed.

Gödel's construction of constructible sets is exactly the same as the construction of the ramified hierarchy except for the following two points.

(1) The construction is continued for all transfinite ordinals, i.e. ramified type α is constructed for every ordinal α. This can easily be done since Gödel's construction takes place inside set theory.

(2) In Gödel's construction, the domain D is also expanded as the construction goes on.

Gödel named the thus constructed sets as the *constructible sets* and denoted the class of all constructible sets by L.

Now the most interesting question here is whether all sets are constructible or not. Denoting the class of all sets by V, the question is expressed by whether $V = L$ holds or not. What Gödel proved is:

(1) L is a model of set theory with $V = L$.

(2) $V = L$ together with axioms of set theory implies the axiom of choice and the generalized continuum hypothesis.

Gödel's result shows that the continuum hypothesis is not a merely technical problem but a problem which is related with a fundamental question on set theory like "Is every set constructible ?"

Modern set theory starts with this work of Gödel. This construction becomes a basic construction in set theory. For example, Paul Cohen proved the independence of the continuum hypothesis in 1963 by using a similar construction. Cohen carried out his construction in an imaginary universe while Gödel did it in the real universe.

As a historical remark, it should be mentioned that Hilbert proposed in his 1926 paper: *Über das Unendliche*, Mathematische Annalen, vol. 95, pp. 161-190, to prove the continuum hypothesis by iterating strictly constructive operations up to constructive ordinals. The combination of Whitehead and Russell's idea and Hilbert's idea is very close to Gödel's construction. Here again we can see a typical Gödel work. He used constructive idea in a nonconstructive way, whereas the others were either simply constructive or simply nonconstructive. In a note Gödel left in the Institute for Advanced Study, he explained that the Generalized Continuum Hypothesis in L is "nothing else but an axiom of reducibility for transfinite order" and carried out his construction and proof *á la* Hilbert.

2.7 The continuum problem and unpublished works

I have discussed his three major works. I should do many other published works, e.g. a decidability result, his interpretation of classical logic in the intuitionistic logic, Gödel-Herbrand's definition of the recursive functions, his relativity work, and so on. But instead of these, let me talk about his 1947 paper : *What is Cantor's Continuum Problem?*, the American Mathematical Monthly, vol. 54, pp. 515-525. There he anticipated that the continuum hypothesis is independent from the present set theory and wrote "its undecidability from the axioms being assumed today can only mean that these axioms do not contain a complete description of the reality" and "the role of the continuum problem in set theory will be to lead to the discovery of new axioms which will make it possible to disprove Cantor's conjecture." In his 1964 supplement of the same paper in Philosophy of Mathematics edited by P. Benacerrof and H. Putnam, Englewood Cliffs, NJ, Prentice Hall, pp. 269-273, he wrote further "despite their remoteness from sense experience, we do have something like a perception also of the object of set theory, as is seen from the fact that axioms force themselves upon us as being true."

His opinion on set theory strongly influenced many set theorists. Search for new axioms especially axioms of strong infinity is now a substantial part of present set theory. The Continuum Problem seems to have been one of Gödel's biggest concerns. In the notes he left in the Institute for Advanced Study, we can see that he worked hard on many problems related to the continuum problem. It is no secret that he made an unsuccessful attempt in 1970 to prove $2^{\aleph_0} = \aleph_2$ by introducing new axioms.

At the end, I also would like to mention Gödel's unpublished works. Many notebooks he left in the Institute show that he was very active during 1940's and 1950's (more precisely till about 1955). Almost all his results of this period are not published. Unfortunately, I cannot discuss these works since they are written in Gabelsberger shorthand.

Gödel's results are basic in all major branches of logic today: model theory, recursive function theory, set theory, proof-theory, and intuitionism. Gödel is indeed the father of the present mathematical logic.

Chapter 3

Hilbert and Gödel

Picasso said something like this, if I remember correctly: In our youth the Academy was all powerful and we shaped ourselves in order to overthrow it. Since there is nothing similarly powerful to overthrow at the present time, one cannot define one's self by opposition to it.

If this is true, then the Academy formed the greatness of Picasso.

Since it appears to me that there is a similar correspondence between Hilbert and Gödel, I would like to explain it. Here, Hilbert's role is that of the Academy and Gödel's is that of Picasso.

To begin to see the relation between Hilbert and Gödel, let us consider, first of all, Gödel's first achievement, that is, the Completeness Theorem.

3.1 The Completeness Theorem

The completeness theorem is a theorem which asserts that the predicate calculus is *complete*. In more detail, it asserts that a proposition which is valid in all mathematical structures is provable in the predicate calculus.

The Completeness Theorem was announced in 1930, when Gödel was twenty four years old, as his dissertation.

It seems to me that the proof of the Completeness Theorem came at a decisive moment when Hilbert and Gödel's scholarly paths intersected.

First of all, the completeness question had been presented as an open problem, two years before Gödel proved it, in "Grundzüge der theoretischen Logik" by Hilbert and Ackermann, published in 1928. Stated in more detail, Hilbert and Ackermann defined the predicate calculus as a formal system for the first time, and, based on that foundation, presented the completeness theorem as an open problem.

There is no doubt that Gödel had read the book by Hilbert and Ackermann, regarded the completeness question as an appropriate problem, considered it and resolved it.

Incidentally, there are some odd facts concerning this. The first one is, as is seen in the previous chapter, the Completeness Theorem is a result which follows immediately from a lemma in an article from 1922 by Skolem.

This fact was discovered by van Heijenoort. When van Heijenoort was about to publish his book "From Frege to Gödel" (Harvard University Press, 1967), he wrote a letter to Gödel and made an inquiry about it.

Gödel's response was as follows: Although it is true that the completeness theorem follows immediately from the article of 1922 by Skolem, Skolem himself was not consciously aware of this, and Skolem's result had remained essentially unknown. For example, in the book by Hilbert-Ackermann, in presenting the completeness question as an open problem, no mention of Skolem is made. Gödel further continued that he did not know of Skolem's work when he proved the Completeness Theorem.

This statement requires a slight correction. A record of the books which Gödel borrowed remains in the library of the Vienna University, and Gödel had borrowed Skolem's article prior to his proof of the Completeness Theorem. Not only that, there are notes in Gödel's handwriting on the pages of Skolem's article. (This is due to Kreisel's investigation.)

It is therefore reasonable to conjecture that Gödel had read Skolem's article but did not clearly remember it. By the time he had forgotten it, he encountered the attractive problem posed by Hilbert-Ackermann, and, while he was struggling to solve the problem, the content of Skolem's article emerged unconsciously to lead him to a solution. There is evidence to substantiate this guess, that is, the method used by Gödel and that of Skolem are essentially the same. There is little difference. If such a guess is correct, then the posing of the problem by Hilbert-Ackermann can be said to be an extremely important factor in the discovery of the Completeness Theorem.

This suggests a very simple explanation for Skolem's inablity to prove the completeness theorem. Namely, at the time of Skolem's 1922 paper, the predicate calculus had not been formalized, and so Skolem could not have attacked the issue simply because he could not conceive the completeness problem itself. It is therefore in no way strange that he could not prove it.

Incidentally, I noticed a strange fact as I read Gödel's article on the Completeness Theorem. In Gödel's article, there is no mention of the open problem posed by Hilbert and Ackermann.

In his article, Gödel takes as his starting point the system of Whitehead and Russell, and states that, in such an approach, the completeness question naturally arises, and he never mentions the proposal by Hilbert and Ackermann. In this article, the only citation of Hilbert and Ackermann is a footnote, in which is written "The predicates and the symbols in this article are due to Hilbert and Ackermann."

I think that was quite unfair to Hilbert. Not only that, the matter is deeper yet. The relationship between Skolem's work and Gödel's Completeness Theorem has been taken up by Hao Wang in his book "From Mathematics to Philosophy" (1974). [1]

Hao Wang writes as follows.

> We do have the striking example of Gödel who possesses firmly held philosophical views which played an essential role in making his fundamental new scientific discoveries, and who is well aware of the importance of his philosophical views for his scientific work. A paper of Skolem in 1922 contains the mathematical core of the proof of the completeness of pure logic. In commenting on the puzzling fact that Skolem failed to draw the interesting conclusion of completeness from his work, Gödel wrote the following paragraphs about the role which his philosophical views played in his work in mathematical logic.

According to this book, Gödel stated that the Completeness Theorem is an almost trivial consequence of the Skolem's work of 1922 but nobody had noticed that fact at the time. He then continued as follows.[2]

> This blindness (or prejudice, or whatever you may call it) of logicians is indeed surprising. But I think the explanation is not hard to find. It lies in a widespread lack, at that time, of the required epistemological attitude toward metamathematics and toward non-finitary reasoning.

Following this, Gödel explains how Hilbert's notion, that all inferences other than those based on the finite standpoint are meaningless, obstructed the resolution of the completeness question. In fact, Gödel did not name Hilbert in the beginning, but it was obvious whom he meant.

---

[1] Routledge & Kegan Paul, Humanities Press, p.8
[2] p.8

To me, this seems to be a one-sided view. For example, Skolem proved various theorems around 1922 by entirely "non-finite" methods, and so Gödel's comment on the fact that Skolem had been unable to obtain the completeness theorem is not at all appropriate. Using common sense, we would conclude that the major reason Skolem had been unable to do what Gödel was later able to do is that only in 1928 did Hilbert and Ackermann formalize the predicate logic and propose the completeness question as an open problem.

My opinion is that Gödel was able to prove the Completeness Theorem and was then mathematically awakened because he was blessed with two kinds of luck.

(1) He had read Skolem's article, but its content did not remain on the surface of his memory.

(2) Hilbert and Ackermann had formalized predicate logic and proposed the completeness question as an open problem at just the right time.

Gödel's activity thereafter was indeed remarkable. In 1931, the year after Gödel had proved the Completeness Theorem, he published the proof of his Incompleteness Theorem, an epoch-making event. Gödel proved the Completeness Theorem when he was twenty four years old. He was rather young, yet it is certain that Gödel was not a prodigy. This becomes apparent if we compare the age of his achievement with the age of Matijasevich when he got his famous result. As we observe Gödel's mathematical achievements thereafter, we can clearly understand how Gödel blossomed and made remarkable progress taking the Completeness Theorem as a starting point.

3.2   The Incompleteness Theorem

Let me refer the reader to my article "Work of Paul Bernays and Kurt Gödel"[3] for an exposition of the Incompleteness Theorem. Here, I would like to mention the following words by Oppenheimer, evaluating Gödel's accomplishment (his Incompleteness Theorem, as a matter of fact).[4]

> (Gödel's great work) illuminated the role of limitations in human understanding in general.

---

[3] Chapter 2 of this volume

[4] Foundations of Mathematics Symposium Papers Commemorating the Sixtieth Birthday of Kurt Gödel, edited by J.J. Bulloff et al., Springer-Verlag,1969: "Greetings to Gödel Symposium from Dr. J.Robert Oppenheimer", p.VIII

The Incompleteness Theorem having been proved, the relation between Hilbert and Gödel took on a darker tone.

The proof of the Incompleteness Theorem caused Hilbert intense sorrow.

In ordinary language, Gödel had shown in this paper that two of Hilbert's fundamental goals were unattainable, at least in the way Hilbert had hoped; one of them was utterly out of the question and the other was virtually so.

Let me state Hilbert's fundamental ideas clearly.

(1) Hilbert thought at least that correct propositions of arithmetic would be provable in an axiomatic system which had been formalized, for example, the system in "Principia Mathematica" by Whitehead and Russell.

(2) Hilbert thought that the consistency proof of a formalized mathematical system could be achieved by naive and combinatorial arguments.

If we take a closer look at the matter, the reality is even worse than one might have suspected.

In the title of Gödel's article on the Incompleteness Theorem, there is the numbering "I." Also, in the last part of the article, it was claimed that a detailed proof of the fact that the so-called consistency proof of an axiomatic system could not be derived within the same system was to be given in the article "II" of the same title as "I." (A brief outline is included in "I.") Why?

The truth of the matter is as follows. Gödel had anticipated fierce resistance from the Hilbert School. Gödel therefore intended to anticipate their reaction and to write the proof in a manner which would convince them over their opposition. Gödel did not need to write "II." As a matter of fact, those in the Hilbert School understood Gödel's article immediately, and did not attempt to refute it at all. Why was that? The fact is that Gödel executed Hilbert's Program more faithfully than anyone else. Those of the Hilbert School, therefore, felt comfortable with the basic conception of Gödel's article.

Let me explain in more detail. Hilbert's basic conception consisted of the following.

(1) One should formalize mathematics.

(2) A formalized mathematical system consists of some symbols and some rules of operations on these symbols, and so it is similar to number theory. Thus one should carry out arithmetic on those objects, which are similar to natural numbers, and prove the consistency of the formalized system.

Here, doing arithmetic on symbols is an original idea of Hilbert.

I believe that the feeling Hilbert had when he began to work on formalism was similar to the feeling he had had when he began to write a report on his work in number theory.

Incidentally, the fundamental method employed by Gödel was arithmetization of a formalized system. That is, it was Gödel who carried out most authentically what Hilbert had claimed and believed. It is an irony of fate that Gödel, who was outside the Hilbert School, executed Hilbert's Program most faithfully and reached conclusions that were the polar opposite of Hilbert's objective.

Consequently, reaction from the Hilbert School (exclusive of Hilbert) was: "What we had intended to do but had neglected, the diligent Gödel did." (This was Bernays' observation.) The result of the Incompleteness Theorem was certainly shocking and unexpected, but the method was accepted without resistence as something that they had known.

Of course, these events saddened Hilbert. In the introduction of "Grundlagen der Mathematik I" (1934) by Hilbert and Bernays, Hilbert writes as follows[5]:

> In consideration of this goal I would like to suggest that the currently fashionable opinion, that the consequence of Gödel's recent results is that the program of my Proof Theory cannot be carried out, is erroneous. These results indeed only show that for the execution of the consistency proof it is necessary to apply the finite viewpoint in a sharper manner than would be necessary for the consideration of a simpler formal system.

Is it my prejudice that I see Hilbert's sorrow all the more in his bluffing? (I have heard that Hilbert had suffered a minor cerebral hemorrhage long before this time, and thereafter he had been unable to be active in mathematics.)

How had Gödel arrived at the Incompleteness Theorem? I will try here to state a guess. I suspect Gödel had realized that the Hilbert School was nothing but a paper tiger when he had proved the Completeness Theorem. In spite of that, he must have been keenly interested in Hilbert's ideas through his experience of proving the Completeness Theorem. Would it not be correct to say that a combination of both moderately positive and

---

[5]Springer-Verlag, The second edition, 1968, p.VII

negative feelings about Hilbert and his ideas guided Gödel to the Incompleteness Theorem?

Incidentally, what is Gödel's view on this matter? In the above mentioned book by Hao Wang[6], Gödel again states as follows.

> How indeed could one think of *expressing* metamathematics *in* the mathematical systems themselves, if the latter are considered to consist of meaningless symbols which acquire some substitute of meaning only *through* metamathematics?

Though it may not be clear from the citation of just one paragraph, it is obviously a criticism of Hilbert's ideas. Gödel is singing the praises of anti-Hilbertian ideas in his Incompleteness Theorem.

And yet, to me this view is not quite appropriate either. It is certain that anti-Hilbertian ideas are seen in the Incompleteness Theorem, but at the same time Hilbertian ideas are also used in it. In fact, there are more of the latter. It is clear that one cannot obtain the Incompleteness Theorem with anti-Hilbertian ideas alone.

3.3   Set theory

The last occasion on which the interrelatedness of Hilbert and Gödel was manifested, came with the famous work of 1938 by Gödel, that is, the consistency proof for the axiom of choice and the continuum hypothesis. Please refer to my article "Work of Paul Bernays and Kurt Gödel" for the content of this work. Here again Gödel used Hilbert's method. Specifically, Hilbert had written an article entitled "Über das Unendliche" in 1926, in which he presented an outline of a possible proof of the continuum hypothesis, and Gödel employed Hilbert's construction exactly as he had given it.

Van Heijenoort, who was mentioned earlier, asked Gödel about this, and in his reply Gödel acknowledged that there was a "remote analogy" between Hilbert's construction and that of Gödel's. Furthermore, Gödel continues as follows.[7]

> There is, however, this great difference that Hilbert considers only strictly constructive definitions and, moreover,

---

[6]p.9
[7]p.369

transfinite iterations of the defining operations only up to constructive ordinals, while I[8] admit, not only quantifiers in the definitions, but also iterations of the defined operations up to *any* ordinal number, no matter whether or how it can be defined.

This problem is brought up again in the above mentioned book by Hao Wang. Here Wang states Gödel's opinion as follows.[9]

With regard to the consistency of the continuum hypothesis, Gödel attributes to a philosophical error Hilbert's failure to attain a definite result from his approach to the continuum problem. The approaches of Gödel and Hilbert are similar in that they both define, in terms of ordinal numbers, a system of functions (or sets) for which the continuum hypothesis is true. The differences are

1. Gödel takes all ordinal numbers as given, while Hilbert attempts to construct them;
2. Hilbert considers only recursively defined functions or sets, while Gödel admits also nonconstructive definitions (by quantification).

In the book by Hao Wang, Gödel's criticism of Hilbert's approach is augmented. If one reads what Gödel told van Heijenoort and Hao Wang, one would get the impression that there is merely a remote and coincidental analogy between the ideas of Hilbert and of Gödel and that they are totally independent. Indeed, in published monographs and papers by Gödel, there is no reference to Hilbert's articles. Is this then the truth?

Gödel does not express anywhere that these two ideas are totally unrelated. He only states the difference of the two constructions as a fact. (We may regard his philosophical arguments as supplementary.) Is it my groundless suspicion that Gödel would not mention Hilbert unless somebody else referred to him, inferring from the case of his Completeness Theorem? Gödel mentioned Hilbert only on those occasions: when the relationship between his Completeness Theorem and Skolem's article was raised, when he replied to a question, and when he criticized Hilbert.

From the discussion above, I conjecture conversely as follows. Gödel

---

[8] Gödel
[9] p.11

carried out his work on the continuum because he became interested in Hilbert's work and studied it. As I wrote in my "Work of Paul Bernays and Kurt Gödel," there are quite a few sets of notes by Gödel in The Institute for Advanced Study in Princeton. Among them, there are a number of manuscripts prepared for lectures on the continuum hypothesis. It seems that Gödel wrote out a new manuscript in detail each time before a lecture. Many of them have clever explanations and are much easier to understand than the existing lecture notes. There is one such set of notes in which he develops his argument with an approach which shows how close his viewpoint is to that of Hilbert. As far as I know, however, there is no record which indicates that Gödel in fact did lecture according to these notes.

Another interesting fact is that, in pursuing this work, Gödel, when he first set to work in 1936 in Vienna, did not use ordinals as something given a priori, but constructed well-orders painstakingly one by one. As a matter of fact, Gödel met Bernays aboard the ship on his way to The Institute in Princeton, and learned from him a system of axiomatic set theory as well as a development of the ordinal numbers. Gödel was thus enabled to present his results in a generally recognized form. (Gödel occupied a first-class cabin, while Bernays was in second-class. At that time, the distinction between first-class passengers and second-class passengers was strict, and so, it is said, the two had to obtain permission from the captain in order to be able to meet and converse on mathematics.)

Viewed in this way, it is at least certain that Gödel's original work was much closer to Hilbert's paper than the published version.

3.4 Hilbert for Gödel

If one judges from publications alone, there is hardly any relationship between Hilbert and Gödel. If there is any, it is manifested negatively. Paradoxically however, I believe all the more that Hilbert's existence had tremendous meaning for Gödel. I can see that Gödel's academic career was molded by the goal of exceeding Hilbert. Having made his contribution to the resolution of the continuum hypothesis at the age of thirty two, Gödel did not do any subsequent work which was comparable to the three works mentioned above. There could be several reasons for that. The fact that there was no longer the challenge to excel Hilbert may have been one such reason.

Gödel took a friendly and infallibly kind approach to me. According to Kreisel, Gödel overrated the two of us because we studied proof-theory. Proof-theory is inseparably connected to Hilbert. Could it not be that this overestimation is related to Gödel's sense of Hilbert's greatness? (Kreisel's explanation is different from this. According to him, Gödel had studied proof-theory eagerly, but was not very successful, and that is why Gödel regarded proof-theorists highly.)

Whenever I met Gödel, he showed a strong interest in Hilbert's Program, consistency proofs and his finite standpoint. This fact strengthens my conviction. Gödel was a person whom I respected and held dear. However, my scholarship essentially belongs to the Hilbert School following Gentzen. That may be another source for my idea.

3.5   Postscript

I would like to add some miscellaneous remarks which are related to the main text, although they are not directly related to its content.

On the book by Hao Wang

This book[10] is notable because it contains various views of Gödel. It is the source of "Gödel aphorisms." Wang explains explicitly which parts were written by Gödel himself and which were written by Wang and approved by Gödel as his opinions. When I met Gödel in November of 1974, Gödel pointed out his own statements and advised me to read only those parts.

In fact, I read Wang's book for the first time in 1980 in order to write "Work of Paul Bernays and Kurt Gödel." Honestly, I felt strange with Gödel's writing on Hilbert, as it gave me an impression of Gödel different from the person whom I had known. I attempted to explain this as follows.

1. It may be that an impression gained from conversation is entirely different from that in print.
2. It may be that one would respond differently when asked something in a letter to be printed later.
3. It may be that, because I was a student of Hilbert's Program, Gödel took a different approach dealing with me than he did in other circumstances.

However, none of these reasons agreed with my recollections of Gödel.

---

[10]From Mathematics to Philosophy by Hao Wang, Routledge & Kegan Paul, 1974

So, I wrote a letter to Kreisel, stating that the impression I got of Gödel as I had read Wang's book seemed quite unlike Gödel, although there was no doubt that many sentences had been written by Gödel himself. I asked Kreisel his opinion of the reason for that.

His opinion was clear cut. He said that Gödel was ill in the seventies, and so he was not in a normal state. However, Kreisel was critical of Wang's book, and expressed his view that it is not an appropriate book from which to gain an appreciation of Gödel's philosophy. In spite of that, he did not seem to think that Gödel's observations in the book seemed unlike him.

It is thus possible that Gödel refrained from criticizing Hilbert in my presence as I belonged to the Hilbert School. From that point of view, the Gödel whom I saw and Gödel in other settings may have been somewhat different.

Notes by Gödel

As a matter of fact, this article was originally intended to present various facts concerning Gödel, and toward the end, observations about the mathematical and psychological relationship of Hilbert and Gödel, which has been discussed so far. Although I began to write with that intention and I have written this much, writing more extensively would involve crossing a threshold. At this moment, I cannot afford to complete my original plan, both emotionally and timewise. So, I have limited the article to its present form, in which the conclusion alone is presented. By the way, as I have mentioned Gödel's notes in the body of this section, I will mention some of my experiences related to these notes.

It started with the following scene: one cold winter morning I went to the Institute in Princeton, recalling that it had been about one year since Gödel had passed away.

Gödel's notes were stored elsewhere, and I went to that place with a secretary. The notes were left neglected in a cardboard box in a corner of a storage room.

I had heard previously that Ms.Underwood had put Gödel's notes into envelopes, without any system. They were in eight or so big Manila envelopes in disorder. I classified and ordered these notes, and then began to record my observations in a room located in the library of the Institute.

Among Gödel's notes, there were some which apparently belonged to his middle school days. They were neatly and carefully written. There were also some which seemed to have been written during his college years. To my surprise, there were many notes which seemed to be manuscripts prepared for talks, such as his lectures at the Institute, Princeton University, Notre Dame University, Harvard University and Yale University. Every one of them was much more detailed and worked out minutely compared with actual lectures and talks. As I compared them with the printed records of lectures, I found that these notes contained a lot of helpful explanations. I was surprised, as I had not known this aspect of Gödel. (When I later told Kreisel this, he completely shared my feelings.) There was a further astonishing fact. Gödel gave lectures, for example, on the consistency proof of the continuum hypothesis, at various places time and again, and each time he prepared a detailed and refined manuscript. Among the numerous notes, those belonging to the following four groups especially drew my interest.

Logic and foundations 1-6,
Mathematical workbooks 1-16,
Results on foundations I-IV,
Max I-XV.

All these were written in "Gabelsberger shorthand", and so I could not read them. In some of them, however, there were titles written in ordinary German, and also some equations as well as some letters which I was able to read, so that I was able to understand vaguely what the subjects were. This heightened my curiosity. I have heard that Gabelsberger shorthand is no longer in use, and there is hardly anybody who can read it. At any rate, I read through whatever I could and looked at it with absorption.

"Logic and Foundations" seemed to be the notes taken from the research of others. There were many names such as Tarski, Mahlo, Sierpinski, Lusin, Souslin, Kuratowski, Herbrand, Gentzen, Rosser and Kleene. I had imagined that Gödel was a studious person, but he was an even harder worker than I had guessed.

However, Gödel's studies appeared to be mainly confined to logic. From what I learned at Vienna University, Gödel read no books and papers except on logic, and, at the examinations, he gave proofs of his own. Among his notes, there are traces of his study of works by Weyl on geometry of numbers and by Siegel on number theory. I suspect Gödel studied these subjects in order to discover some undecidable arithmetic propositions, which had been proved to exist by his Incompleteness Theorem. (Both Weyl and Siegel were present in Princeton and were German speaking, and I suppose Gödel conversed with them and studied their theories.) Judging from Gödel's notes, however, Gödel did not get too deeply into these areas. "Mathematical workbooks" and "Results on foundations" appeared to be notes of his own results. "Mathematical workbooks" seemed to have been written every day as a spontaneous record of his thoughts. "Results on foundations" would appear to be what he wrote when he reached specific conclusions. There were overlapping paragraphs, some of which I thought had been copied from one to the other. "Mathematical workbooks" dated from his Vienna days to his Princeton days. For example, "Mathematical workbook 13" had the date September 3, 1941. "Results on foundations" evidently was started after his arrival in America. At the beginning of "Results on foundations I" it was written[11] "I start in America, March 1940." I could feel the Gödel's joy. He loved America, especially Princeton. I could only roughly guess the content in these notes from the titles, but I felt there were many subjects concerning the "continuum" and the "scale", which I had heard in our private discussion. However important they might have been, they were useless as I could not read them. Thus I took brief notes and some Xerox copies for later reference. That was as much as I could do.

"Max" was continued from his days in Vienna to Princeton. "Max VIII-XIV" was from 1942 to 1955, and "Max XV" lasted from 1955 to his last days. It ended abruptly. It appeared to me that in these notes were written whatever came to his mind in a diary-like style on philosophy, psychology, theology, foundations of mathematics and mathematics. Most of the content was philosophy, but in "Max V" (1943-44) there is a fair amount of foundations of mathematics and in "Max VIII" (1942) and "Max XV" (1946-55) some foundations of mathematics and mathematics.

"Max XV", the last one, contains mostly philosophy, but a little bit of foundations of mathematics.

I had to put it all in order and extract what I could of the notes within a short half day, and I felt rushed. As I read scratches here and there in the notes, many recollections came back to me, which made me miss the old days. What was more, I keenly felt the void caused by Gödel's absence and his unhappy later years.

Additional note

The Hilbert School had been a dominant academic school until Gödel appeared. Hilbert was unquestionably a strong influence on Gödel. However, I do not know if the psychological influence that I conjectured in the main part of the text was real. I cannot rigorously demonstrate such an influence.

Although I have written these things as I thought such a possibility was

---

[11] more or less

interesting, it does not mean that I have complete confidence in them. I need to write what I believe, otherwise it will be difficult for me to express my thoughts.

Chapter 4

Short Biographies of Logicians

When I read "The Double Helix" by James D. Watson, the famous DNA researcher, I found references to my good friend Kreisel, which pleased me. It is said that Watson wrote so frankly that now two persons mentioned in his book have charged him with libel. (According to gossips, the observations are not defamatory, since all incidents related are factual.) Although this chapter is entitled "Short Biographies of Logicians," it is in fact a personal recollection of friendships, and its true purpose is a collection of rambling observations. I would ask the reader's indulgence.

4.1 Kurt Gödel

4.1.1 *The extraordinary works of Kurt Gödel*

From the perspective of logicians, Gödel seems godlike. He was shy and avoided company, and that makes us feel he is even closer to God. An amazing collection of brilliant ideas in logic and set theory originated in Gödel's work.

Let me start with a brief vita of Gödel. Gödel was born on April 28th, 1906, in Brün (present Brno) of Czechoslovakia, at the time part of the Austro-Hungarian Empire. He studied mathematics and physics at the Vienna University, was granted the Doctorate of Philosophy in 1930, served as a Privatdozent during 1933 to 1938 at the Vienna University, and visited at the Institute for Advanced Study in Princeton as a researcher several times. Thereafter he worked as a researcher at the Institute from 1940 throughout 1953, and was promoted to professor in 1953. He acquired U.S. citizenship in 1948. (There is a famous story that, while preparing for the examination of the American Constitution in order to acquire citizenship,

Gödel told Einstein: "It is awkward that the American Constitution is not consistent.") Gödel was awarded an Honorary Doctorate of Literature by Yale University in 1951 and an Honorary Doctorate of Science by Harvard University in 1952. He was elected to National Academy of Sciences and also received the first Einstein Award in 1951.

The greatness of Gödel's work is radiantly obvious to logicians. Yet, it is not easy to explain in ordinary language. Consequently, I fear it might be mistaken for mere good work. For now, I would like to cite the late J. Robert Oppenheimer and John von Neumann [1]. (They both understood Gödel quite well as his colleagues at the Institute and as distinguished scholars of physics and mathematics.) Oppenheimer wrote as follows[2]:

(Gödel's great work) has not only immeasurably deepened and enriched the understanding of the logical structure of so much abstract and mathematical argument, but illuminated the role of limitation in human understanding in general.

Von Neumann wrote as follows, when Gödel received the Einstein Award in March of 1951[3]:

Kurt Gödel's achievement in modern logic is singular and monumental - indeed it is more than a monument, it is a landmark which will remain visible far in space and time. Whether anything comparable to it has occurred in the logic of modern times may be debated. In any case, the conceivable proxima are very, very few. The subject of logic has certainly completely changed its nature and possibilities with Gödel's achievement.

Gödel's name is associated with many important achievements in detail, and with two absolutely decisive ones. The occasion is such that I think I should only talk about the two latter.

The nature of the first one is easy to indicate, although its exact technical character and execution escape an adequate characterization without the specialized and rather intricate techniques of formal logic.

Gödel was the first man to demonstrate that certain mathematical theorems can neither be proved nor disproved with the accepted, rigorous methods of mathematics. In other words, he demonstrated

---

[1] Foundations of Mathematics, Springer-Verlag, 1969
[2] p.VIII
[3] ibid. pp.IX-X

the existence of *undecidable* mathematical propositions. He proved furthermore that a very important specific proposition belonged to this class of undecidable problems : The question, as to whether mathematics is free of inner contradictions. The result is remarkable in its quasi-paradoxical 'self- denial' : It will never be possible to acquire *with mathematical means* the certainty that mathematics does not contain contradictions. It must be emphasized that the important point is, that this is not a philosophical principle or a plausible intellectual attitude, but the result of a rigorous mathematical proof of an extremely sophisticated kind.

The formulation that I gave above has coarsened the result and obliterated some of the fine points of its rigorous formulation, but if one is to state the theorem without having recourse to the difficult technical language of formal logic this is, I think, the best approximation that one can achieve.

Gödel actually proved this theorem, not with respect to mathematics only, but for all systems which permit a formalization, that is a rigorous and exhaustive description, in terms of modern logic: For no such system can its freedom from inner contradiction be demonstrated with the means of the system itself.

Gödel's second decisive result can only be stated in the terminology of formal logic and of an important but rather abstruse modern mathematical discipline: Set theory. Two surmised theorems of set theory, or rather two principles, the so-called 'Principle of Choice' and the so-called 'Continuum Hypothesis' resisted for about 50 years all attempts of demonstration. Gödel proved that neither of the two can be disproved with mathematical means. For one of them we know that it can not be proved either, for the other the same seems likely, although it does not seem likely that a lesser man than Gödel will be able to prove this.

I will not attempt a detailed evaluation of these achievements, I will limit myself to iterating: In the history of logic, they are entirely singular. No indemonstrability within mathematics proper had ever been rigorously established before Gödel. The subject of logic will never again be the same.

4.1.2   *What is the Incompleteness Theorem?*

In the words of von Neumann, the first of the two decisive results is what is usually called "Gödel's Incompleteness Theorem," which was published in 1931. The central problem in logic of the time was the consistency problem proposed by the Hilbert School. Von Neumann was a gifted young disciple of that School who was then giving lectures on the consistency proof. But one day as he entered the classroom he said "Recently Gödel has proved the impossibility of the consistency proof. I have examined his proof, and am convinced it is correct. So, I shall terminate as of this lecture." Following this, it is said, he indeed quit. [4] His last sentence, "The subject of logic, ..." must have meant that the consistency problem pursued by the Hilbert School would never again be a central issue of logic.

Gödel himself states in the paper presenting the Incompleteness Theorem that this result does not by any means signal the failure of Hilbert's Program. Gödel's attitude was cautious. Indeed in 1958 he developed a method of consistency proof for number theory which seemed quite different from that of Gentzen's, although the two methods are deeply related in substance. Nevertheless, it is an indisputable fact that, because of the incompleteness result, the consistency problem fell out of the spotlight and became a problem which would draw the attention of only a rare few.

The comments by von Neumann above, explicitly characterized his personal view as well as that of the era of 1951. For example, what he calls mathematics should be interpreted as grounded in Zermelo-Fraenkel set theory. It expresses von Neumann's belief that mathematics is based on this Zermelo-Fraenkel set theory as a fixed foundation. Also, it is well-known that the independence of the continuum hypothesis and the axiom of choice were later established by Paul Cohen. (Contrary to von Neumann's statement, the issue of the independence of the axiom of choice had not been completely resolved at that time.) Such trifles aside however, von Neumann's review seems appropriate.

Let me explain Gödel's work in greater detail.

As you may know, in mathematics, many notions of various kinds have to be defined. Upon logical analysis of such notions, they are composed of the logical notions $\neg$ (not), $\wedge$ (and), $\vee$ (or), $\rightarrow$ (implies), $\forall$ (for all), and $\exists$ (exists), and some basic notions which are specific to the mathematics in

---

[4] Author's note: This may not be accurate. Circumstances concerning von Neumann's reaction and Gödel's incompletness results are described in detail in page 70 of "Logical Dilemmas" by John W. Dawson, Jr. published by A K Peter, 1997.

question ($=, <, +, \times$ for number theory, and $=, \epsilon$ for set theory). We may understand that a study of mathematics in terms of analysis like this made gradual progress since the last century[5], and was completed around the beginning of this century[6] (Leibniz, de Morgan, Peirce, Schröder, Frege, Peano, Whitehead, Russell). Through such investigations, it turned out that a mathematical proof can be composed starting from a few mathematical axioms (concerning basic notions of a mathematical theory) and repeated use of only a few simple rules characterizing the logical symbols as above. This was a great success of mathematical logic.

A natural question arises then.

Could there be some new logical rules other than these finitely many logical rules we have so far used? In other words, Is our system of logical rules complete?

Gödel's first paper (1930) gave a complete solution to this problem, that is, it proved the following.

> Our logic is complete. Namely, if a proposition P is unprovable in our logic from a set of axioms Γ, then we can construct a model which simultaneously satisfies Γ and $\neg P$.

This theorem is called the Gödel's Completeness Theorem.

The next paper by Gödel (published in 1931) is the article on the Incompleteness Theorem, whose exposition was cited earlier. In order to explain the meaning of this article, I need to say something about set theory.

As you may know, set theory, which had been developed by Georg Cantor towards the end of the nineteenth century, became the central stream in the torrent of modern mathematics. Although paradoxes such as those of Russell and Burali-Forti had given rise to serious foundational problems at one time, a stable set theory was eventually completed by the efforts of Zermelo, Fraenkel, von Neumann, Bernays and others. Now, there is no question that all fields of modern mathematics are included within this set theory. [7] At present, quite a few mathematicians believe consciously or unconsciously that modern mathematics equals ZF set theory[8]. (For this reason, some people believe, upon hearing that the continuum hypothesis is independent of ZF set theory, mathematicians can never choose a position concerning the continuum hypothesis, and as a result get lost in wild philo-

---

[5] 19th century

[6] 20th century

[7] In the sense that they are logically grounded in and derivable from this theory.

[8] Zermelo-Fraenkel set theory

sophical fantasies. The fact is, it only means that some axioms are missing in ZF set theory, and an interesting and positive problem of which axioms to add awaits us. However, this is a circumstance which became clear when the implications of Gödel's Incompleteness Theorem were absorbed. It had been natural to see things as above until 1930. Many mathematicians had consequently believed that all mathematical problems could be solved, at least in principle, in ZF set theory. Gödel's theorem showed this assumption to be false.

Namely, if ZF set theory does not yield a contradiction, then there is a problem which cannot be decided within ZF set theory. This problem is a simple one in the sense that it can be described in terms of $=, +, \cdot$ and logical symbols alone. The meaning of its content is the consistency of ZF set theory. In general, given any formal system, if the system can be fully described and is consistent, and number theory can be carried out within the system, then a sentence which states "the system is consistent" can be expressed in the language of the system, yet this sentence is undecidable within the system, that is, neither the sentence itself nor its negation can be proved in the system.

This theorem, as von Neumann stated, shows that, not only is the present ZF set theory incomplete, but also it is impossible to obtain a univeral system, no matter how one improves ZF set theory, in which, in principle, all propositions are decidable. Furthermore, such an undecidable problem is not something that is unrelated to modern mathematics, but can be expressed in the language of number theory. This fact makes the situation all the more interesting.

Although this theorem, as von Neumann claimed, shows something of the limits to human intelligence, it also produced a serious dilemma in the history of logic. Russell's paradox, found in the early development of set theory urged upon logicians serious reconsideration of the foundations of mathematics, but Hilbert proposed his formalism and tried to save mathematics from the crisis by proving the consistency of various mathematical systems. As I already mentioned, this theorem of Gödel presented a fact which derailed Hilbert's plan. (In the field of consistency problems, there appeared a proof of the consistency of number theory by Gentzen in 1936, but we cannot say that there have been many glorious subsequent achievements in this area.)

4.1.3   *The starting point of modern set theory*

Gödel's other important achievement belongs to the field of set theory. (I do not mean that other works of his are not important. There are not many articles by Gödel; there are about thirty. There are many other important contributions, but, after all, the three mentioned here are the most outstanding among them.)

Let me explain set theory a bit. There are two central principles in set theory, the principles of set construction.

(1) The principle of subset construction - Given a domain D, it creates subsets of D. (D is assumed to have already been constructed and to be unproblematic. In many cases, the domain consisting of all natural numbers can be regarded as such.)

As a special case of this principle, one can construct "the set of all elements x of D which satisfy a property $P(x)$" when D is a set. This is a subset of D. The collection of all such subsets of D is called the power set of D, and so this can be called the principle of creating the power set.

(2) The principle of ordinal construction - One may consider this as a principle of transfinite iteration. Explained more concretely, it is a principle which allows the continuation of the following constructions as long as possible. Starting with the empty set 0 (The empty set is also denoted by ϕ, but here I write it as 0.), name the "set which consists of 0 alone" as 1, the "set which consists of 0 and 1 alone" as 2, the "set which consists of 0, 1 and 2 alone" as 3, and so and so forth, and thus construct all the natural numbers. Then name the "set of all natural numbers" as ω, the "set of all natural numbers and ω" as $\omega + 1$, the "set of all natural numbers, ω and $\omega + 1$" as $\omega + 2$, and so on and so forth.

$\omega + \omega$ will denote the "set consisting of all natural numbers and of sets of the form $\omega + n$ for all natural number n," and $\omega + \omega + 1$ will denote the "set consisting of all the elements of $\omega + \omega$ and of $\omega + \omega$." [9] We may write $\omega + \omega$ as $\omega 2$ and $\omega + \omega + \omega$ as $\omega 3$. Similarly ω^2 will denote the "set consisting of all sets of the form $\omega n + m$ with n and m natural numbers," and $\omega^2 + 1$ will denote the "set consisting of all sets which are elements of ω^2 as well as ω^2." There is really no end. After $\omega^2, \omega^3, \cdots$, there comes a set denoted by ω^ω. However, no matter how far we go with such iteration, we can obtain only countable ordinal numbers. After all these countable ordinals, there comes the first uncountable ordinal $\aleph_1$, and, considering that this $\aleph_1$ is regarded as a truly small ordinal among all the ordinals, the story of ordinals makes

---

[9] $\omega + \omega = \{0, 1, 2, 3, \cdots, \omega, \omega + 1, \omega + 2, \cdots\}$

us feel faint.

I have written above that we continue the construction of ordinals "as long as possible." A big problem of modern set theory is how to express this "as long as possible." Fraenkel's axiom of replacement is a partial expression of this principle "to continue creation of ordinals as long as possible," but in contemporary set theory one can think of arbitrarily strong axioms of such creation. The problem is how to acquire an essentially new axiom.

It is no exaggeration to say that these two principles (the axioms of the power set and of ordinals) are all that matter in set theory. Namely, every set can be obtained by starting with the empty set and by transfinite iteration of application of the principle of constructing a power set along with construction of ordinals. We can thus say that the study of the axioms of set theory will be complete if we can understand all about constructions of ordinals and power sets.

Let me return to Gödel's works. Our first principle–the principle of power sets– in fact contains various difficulties. Let N denote the domain of all natural numbers, and let us assume that we know all about N (that is, about the definition of $0, 1, 2, 3, \cdots$ and the definition of $=, <, +, \cdot$). Then, let us adopt the method of defining subsets of N by which we construct an unknown set from known sets one by one with confidence, instead of assuming that set theory is an axiom system from heaven. (This was first asserted by Russell.) When we try to define a subset of N as $\{x|x \in N \& P(x)\}$ [10] according to this method, the property of $P(x)$ must be sufficiently well defined.

Let us consider now a case where the property $P(x)$ is not sufficiently well defined. What if a notion such as "for all subsets X of N" or "for some subset X of N" is involved in $P(x)$? If we abide by our principle to "construct an unknown set from known sets one by one with confidence," then all the notions contained in $P(x)$ have to be clearly defined in order that $\{x|P(x)\}$ be clearly defined (or constructed). For that, the notions such as "for all subsets of N" and "there exists a subset of N" have to be clearly defined. For such a notion to be clearly defined, all the subsets of N have to be already clearly defined. Hence, it does not make sense to define anew a set $\{x|x \in N \& P(x)\}$ by this definition as a special case of a subset of N. This is Russell's so-called circular reasoning.

In general, we can identify a real number with a set of natural numbers,

---

[10] "the set of x in N satisfying $P(x)$

and so the operation of taking the supremum of a set of real numbers in fact corresponds to the notion of "for all subsets X of N" or "for some subset X of N" as mentioned above. For this reason, analysis contains essentially circular reasoning, and hence its grounding is an interesting problem from the view point of the foundations of mathematics.

Now, if we consider a proposition $P(x)$ which consists of only $\forall y \in N$[11], $\exists y \in N$[12], $\neg$ (not), $\wedge$ (and), $\vee$ (or), $\rightarrow$ (implies), $=, <, +, \cdot$, then for a natural number x the meaning of $P(x)$ is clearly defined from some known notions. (Incidentally, a proposition such as this $P(x)$ is said to be arithmetic.) The set $\{x | x \in N \& P(x)\}$ can therefore be also said to be clearly defined. Let us call such a set an arithmetic set. If we denote the collection of all arithmetic sets by T_1, then we may say that T_1 is well defined. With regards to N and T_1, aside from $=, <, +, \cdot$ on the elements of N, new relations $n \in A$ between an element n of N and an element A of T_1 and $A = B$ between two elements A and B of T_1 can be defined. This new domain T_1 together with $=, <, +, \cdot, \in$ can be said to be well defined. Thus, relations such as $\forall A \in T_1, \exists A \in T_1$ and $n \in A$ for $n \in N$ and $A \in T_1$ can be said to be well defined.

Suppose a proposition $Q(A)$ is composed of $\forall n \in N, \exists n \in N, \forall B \in T_1, \exists B \in T_1, +, \cdot, n \in A$. Let us call $Q(A)$ a predicative proposition with respect to T_1. Then we may say that the set $\{A | A \in T_1 \& Q(A)\}$ (This is called a predicative set with respect to T_1.) is well defined. If we let T_2 be the collection of all the predicative sets over T_1, then we may again consider that T_2 is well defined.

We can repeat this consideration over and over. When an ordinal number α is given, we denote the object created by repeating such a construction α times by T_α. Any set a belonging to T_α can be regarded as being constructed in terms of the principle of constructing ordinals–the second principle– and a universal principle which defines an unknown object from some known objects with confidence, without resorting to the first principle. Gödel named such a set a "constructible set." If we take a sufficiently large α, then we can claim that T_α reflects the universe of sets. Gödel clearly defined the notion of a constructible set in ZF set theory, and then proved the following.

"Let L denote the class of all constructible sets. L is a model of ZF set theory as well as of the axiom of choice and the generalized continuum hypothesis."

---

[11] "for all y in N"

[12] "there exists y in N"

This theorem is certainly important in that it proves the consistency of the axiom of choice and the generalized continuum hypothesis relative to ZF set theory, but it is even more interesting for the following reason. It proposed the important notion of L and showed that L satisfies the axiom of choice and the continuum hypothesis.

Indeed, what Gödel demonstrated is the consistency of the assumption that all sets are constructible. Since this fact implies that set theory can be established without the first principle, it has philosophical interest. (It is erroneous to believe that the foundations of analysis have been completed with this fact, as the second principle in set theory is by far a more transcendental principle than the construction of subsets in analysis.)

The question "Are all sets constructible?" is a serious problem in modern set theory, and the opinions of specialists are divided evenly. By the way, Paul Cohen proved that this proposition is independent of ZF set theory, and Gödel asserted that this proposition is in fact false. I will take up this matter again later. At any rate, in fact Gödel's investigation as above can be called the starting point of modern set theory.

At present, set theory is so prosperous that it is said the major subject of logic is set theory. Young distinguished set theorists such as Solovay, Silver, Martin and Kunen are being produced one after another. It should not be long before set theory becomes one of the lions in modern mathematics. Well, I might say it already has. Most of contemporary set theory can be said to be the extension of Gödel's work as described above.

For example, many of the specialists view the notion of "forcing" [13], which is now a classic, as a slight extension of Gödel's above mentioned idea. (What Gödel executed on real sets, Cohen executed in the world of dreams!)

4.1.4 *Belief in set theory*

Having completed his great work, Gödel hardly did any mathematics subsequently. (He announced the results above in 1938 and the outline of the proof in 1939, and then gave lectures at the Institute for Advanced Study on its details. The lecture notes were published by Princeton University in 1940.) There were only two publications since then, one on relativity (1950) and one on the consistency of number theory (1958). However, his influence on logic, especially on set theory, was extremely strong, and it is no exaggeration to say that his influence was the propulsive power behind

---

[13] Cohen's method of proving independence results relative to ZF set theory

the progress of modern set theory.

During my stay at the Institute in Princeton, I was often asked by researchers from other places "What is Gödel's opinion about this?" relative to various problems. On some occasions when I communicated Gödel's opinion to a questioner, another person listening in would laugh and say "That settles the problem!" Gödel's influence was that strong. Here I would like to introduce Gödel's belief with regards to set theory (It is too strong to be called an opinion.).

Gödel's belief was that set theory is not at all a mathematical game which abides by artificially decided rules. He believed that the collection of all sets is existentially clear and real, and that all problems in set theory are to be resolved one day either affirmatively or negatively.

Although the continuum hypothesis was proved to be independent of ZF set theory by Cohen, it simply means that ZF set theory lacks some axioms. The problem is how to discover the necessary new axioms. Of course, it is impossible, due to the Incompleteness Theorem, to create a universal set theory in which all problems can be solved, but he believed it to be possible that, when a particular problem is presented, one can discover an appropriate new axiom to resolve it. He believed that, since the totality of all sets has a clear and real existence, such a possibility is not only correct in principle but also realistic.

For example, he asserted that the continuum hypothesis is false (This implies the existence of a non-constructible set, since the continuum hypothesis holds if all sets are constructible.), and gave various grounds for this belief. Most of them involve the claim that the assumption of the continuum hypothesis induces unnatural phenomena. I once asked him: "Do you decide if a new axiom is valid by looking at its consequences?" His answer was "Not necessarily. However, the case of the continuum hypothesis is similar to the situation in which one feels very odd when just one note is out of tune in a beautiful and simple melody. That is why I believe it is false." Therefore, these phenomena cannot be decisive factors in denying the continuum hypothesis, and he himself eagerly awaited the discovery of a new and incontrovertible axiom which would deny the continuum hypothesis.

One should not take this opinion of his lightly. It comes from a profound philosophical belief in the beautiful harmony of the laws of the universe. When Gödel's belief in the negation of the continuum hypothesis became widely known, many specialists accepted it and imagined that then the cardinality of the continuum would be extremely large. Surprisingly, however,

Gödel's opinion was that the cardinality of the continuum would not be so large, and he said, although he was not certain, it might be $\aleph_2$.

4.1.5 *Gödel's fundamental philosophy*

Now, let us return to the standpoint of Gödel's fundamental philosophy. Since wording is important for a philosophical argument, I will cite Gödel's writing from "Philosophy of Mathematics: selected readings, 2nd edition," edited by P. Benacerraf and H. Putnam, CUP, pp.482-485. Gödel sometimes refers to some relevant pages, which I quote without remarks. Also, I will omit original footnotes.

In the following, an inaccessible number is a regular cardinal **n** which satisfies $\alpha < \mathbf{n} \to 2^\alpha < \mathbf{n}$. (There is general agreement in modern set theory that the axiom claiming the existence of an inaccessible cardinal number is correct.)

Now Gödel's writing.

"$\cdots$ it has been suggested that, in case Cantor's continuum problem should turn out to be undecidable from the accepted axioms of set theory, the question of its truth would lose its meaning, exactly as the question of the truth of Euclid's fifth postulate by the proof of the consistency of non-euclidean geometry became meaningless for the mathematician. I therefore would like to point out that the situation in set theory is very different from that in geometry, both from the mathematical and from the epistemological point of view.

In the case of the axiom of the existence of inaccessible numbers, e.g., (which can be proved to be undecidable from the von Neumann- Bernays axioms of set theory provided that it is consistent with them) there is a striking asymmetry, mathematically, between the system in which it is asserted and the one in which it is negated.

Namely, the latter (but not the former) has a model which can be defined and proved to be a model in the original (unextended) system. This means that the former is an extension in a much stronger sense. A closely related fact is that the assertion (but not the negation) of the axiom implies new theorems about integers (the individual instances of which can be verified by computation). [14] So the criterion of truth on p.[476] is satisfied, to some extent, for the assertion, but not for the negation. Briefly speaking, only the assertion yields a "fruitful" extension, while the negation is sterile

---

[14]Author's Note: A theorem in this context is of the form $\forall x \in N.P(x)$ and "each instance" means $P(n)$ for each natural number n.

outside its own very limited domain. Cantor's continuum hypothesis, too, can be shown to be sterile for number theory and to be true in a model constructible in the original system, whereas for some other assumption about the power of the continuum this perhaps is not so. On the other hand neither one of those asymmetries applies to Euclid's fifth postulate. To be more precise, both it and its negation are extensions in the weak sense.

As far as the epistemological situation is concerned, it is to be said that by a proof of undecidability a question loses its meaning only if the system of axioms under consideration is interpreted as a hypothetico-deductive system; i.e., if the meanings of the primitive terms are left undetermined. In geometry, e.g., the question as to whether Eudlid's fifth postulate is true retains its meaning if the primitive terms are taken in a definite sense, i.e., as referring to the behavior of rigid bodies, rays of light, etc. The situation in set theory is similar, the difference is only that, in geometry, the meaning usually adopted today refers to physics rather than to mathematical intuition and that, therefore, a decision falls outside the range of mathematics. On the other hand, the objects of transfinite set theory, conceived in the manner explained on pp.[474-5] and in footnote 11, clearly do not belong to the physical world and even their indirect connection with physical experience is very loose (owing primarily to the fact that set-theoretical concepts play only a minor rule in the physical theories of today).

But, despite their remoteness from sense experience, we do have something like a perception also of the objects of set theory, as is seen from the fact that the axioms force themselves upon us as being true. I don't see any reason why we should have less confidence in this kind of perception, i.e., in mathematical intuition, than in sense perception, which induces us to build up physical theories and to expect that future sense perceptions will agree with them and, moreover, to believe that a question not decidable now has meaning and may be decided in the future. The set-theoretical paradoxes are hardly any more troublesome for mathematics than deceptions of the senses are for physics. That new mathematical intuitions leading to a decision of such problems as Cantor's continuum hypothesis are prerfectly possible was pointed out earlier (pp.[476-7]).

It should be noted that mathematical intuition need not be conceived of as a faculty giving *immediate* knowledge of the objects concerned. Rather it seems that, as in the case of physical experience, we *form* our ideas also of those objects on the basis of something else which *is* immediately given. Only this something else here is *not*, or not primarily, sensory. That something besides sensations actually has immediacy follows (independently of mathematics) from the fact that even our ideas referring to physical objects contain constituents qualitatively different from sensations or mere combinations of sensations, e.g., the idea of object itself, whereas, on the other hand, by thinking, we cannot create any qualitatively new elements, but only reproduce and combine those that are given. Evidently the "given" underlying mathematics is closely related to the abstract elements contained in our empirical ideas. It by no means follows, however, that the data of this second kind, because they cannot be associated with actions of certain things upon our sense organs, are something purely subjective, as Kant asserted. Rather, they too may represent an aspect of objective reality, but, as opposed to sensory data, their presence in us may be due to another kind of relationship between ourselves and reality.

However, the question of the objective existence of the objects of mathematical intuition (which, incidentally, is an exact replica of the question of the objective existence of the outer world) is not decisive for the problem under discussion here. The mere psychological fact of the existence of an intuition which is sufficiently clear to produce the axioms of set theory and an open series of extensions of them suffices to give meaning to the question of the truth or falsity of propositions like Cantor's continuum hypothesis. What, however, perhaps more than anything else, justifies the acceptance of this criterion of truth in set theory is the fact that continued appeals to mathematical intuition are necessary not only for obtaining unambiguous answers to the questions of transfinite set theory, but also for the solution of the problems of finitary number theory (of the type of Goldbach's conjecture), where the meaningfulness and unambiguousness of the concepts entering into them can hardly be doubted. This follows from the fact that for every axiomatic system there are infinitely many undecidable propositions of this type.

It was pointed out earlier that, besides mathematical intuition, there exists another priority that gives weight to the truth of mathematical axioms, namely their fruitfulness in mathematics and, one may add, possibly also in physics. This criterion, however, though it may become decisive in the future, cannot yet be applied to the specifically set-theoretical axioms (such as those referring to large cardinal numbers), because very little is known about their consequences in other fields. The simplest case of an application of the criterion under discussion arises when some set-theoretical axiom has number-theoretical consequences verifiable by computation up to any given integer. On the basis of what is known today, however, it is not possible to make a convincing argument of the truth of any set-theoretical axiom in this manner."[15]

4.1.6 *A gentle and fair minded scholar*

I have spent much space in explaining Gödel's works and philosophy. I would like to add here some personal information.

His build was almost the same as mine, and I am rather small even as a Japanese. This made me feel close to him. Although he had come to America more than thirty years earlier, his English had a strong German accent. He seemed to have an extreme dislike of meeting people for the first time. But I felt that he was rather amiable to close friends.

I talked with him only about mathematics, but there was a logician who rarely discussed mathematics with him but talked on the phone with him for hours at a time about Vietnam. Gödel was a great friend of Einstein's and was on good terms with Bernays. Is this in part because he could converse with them in German? During 1959 to 1960 when Bernays was at the Institute in Princeton, Gödel talked with Bernays for many hours, and I once asked Bernays what in the world they discussed. Bernays answered that they talked about a variety of subjects including a particular store in a European town. It seems that mathematics was not their only subject. I heard that Gödel sent his wife to Vienna for vacation every summer, but he did not accompany her. (On such occasions, I saw him alone eating vegetables quietly in a corner of the cafeteria in the Institute.) When Mrs. André Weil asked "Why don't you go with her?" he answered he was unable to travel as he had stomach trouble. As I come to think of it, his paper on the Incompleteness Theorem is numbered "I" and the details of the latter part of the proof were to be carried out in "II," but "II" has never been published. I heard that it was because his stomach trouble was quite serious, causing him nightly insomnia and making him feeble, so that he was unable to write the paper. (The part which was to be included in "II" would indeed cause stomach trouble.)

Although he was famous for not attending parties at all, he occasionally did come. For what reason I do not know. Even then I would say he was not sociable. In 1966, the new director of the Institute, Dr. Carl Kaysen hosted his first big party at the Director's official residence (an excellent house with authentic paintings by Manet and others). Somehow Gödel was present at the party without his wife, and he appeared bewildered by the

---

[15]The end of citation from Gödel's writing

enormous crowd. (He must have thought it would be a small gathering attended only by faculty.) As soon as he found me, he approached me and chatted for about an hour and a half, and then left saying "I must go now." During that time, only André Weil and his wife as well as Kochen, a logician, joined our conversation. I had spoken with him the day before, and I could talk about nothing but mathematics, and so I cannot but guess that he really disliked meeting strangers.

I have never known anyone else so considerate, fair, selfless, deep and broad-minded as Gödel. (In most cases one's inclination is to promote a philosophy that is most convenient to one's own opinions, but he had no such tendencies at all. He, who had proved the Incompleteness Theorem, was most interested in consistency proofs. He, who had proposed the interesting notion of constructibility and proved that the axiom "all sets are constructible" yields many amazing results, believed in the existence of non-constructible sets. I do not know of anybody else who has so deserved to be called a Scholar as he did.)

4.1.7 *Gödel and myself*

Nevertheless, I was sometimes perplexed that Gödel was so different from ordinary human beings. Once I needed a recommendation letter to send to a certain university. I asked him to write one and he consented to do so with pleasure. But he did not write it for some time. The university needed it in a hurry, and the person in charge became impatient. He phoned me and pressed me to ask Gödel to write it as soon as possible. I reluctantly went to his office to encourage him. Stepping into his office (It was a new, large and beautiful office with a wonderful view of a pond and woods, but he kept the curtain closed. He must have disliked to be seen from outside.), I found him with my papers spread all over the desk. He told me that as he had not read my papers in detail yet, he was unable to write my recommendation letter.

Some time later, a telegram was sent directly to Gödel and I received a telephone call from the University, urging me to expedite the matter. So, I went to Gödel and asked him to please write the recommendation, telling him that a superficial one would do. He agreed reluctantly to write one, adding that he could not say anything definite as he had not made a careful reading of my papers. (Gödel, who claimed not to have read my papers carefully, seemed to understand thoroughly. At least he grasped them far more accurately than Kreisel, who writes reviews of them.)

There was another related incident. I did some modest work on the consistency problem in the spring of 1963. It was the fall of that year when I wrote it up as an article. (Of course I sent a copy to Gödel immediately.) When I saw him in the summer of 1964, I asked if it would be appropriate for me to submit the article to "Annals of Mathematics." He said it would. I then revised the introduction and other parts, making a final manuscript, and sent it to him. When I met with him in the summer of 1965, I told him I would entrust the article to him. So, during my visit to the Institute in the summer of 1966, I assumed the refereeing work and other business had been completed and the article was about to be printed. When I asked him about it, he replied "Have you found a referee?" I was surprised and answered "I assumed you would find a referee." He told me that anybody would do, and suggested some possible referees. I then asked one of them to review my article, received a reply from him and there was some further interaction. It was in December of 1966 when Gödel finally submitted my paper to "Annals." (After that, "Annals" officially asked the person I had addressed earlier to serve as a referee.)

Gödel then asked me why it was that the content of the article in question had not been changed, notwithstanding that three years had passed since he had first received it. Somehow I felt a pang of conscience. I felt very much relieved when in the fall of 1967 the article was published. But, according to what I heard later from an informant, "Annals" had a policy of not carrying articles in pure logic which were unrelated to other fields of mathematics. That may have been the reason why Gödel was hesitant.

One of the pleasures of living in America for me was meeting Professor Gödel from time to time and discussing mathematics. He always said " I recently have been studying philosophy and so I cannot follow what you do ····." And yet, to my surprise, he was well acquainted with recent articles and had deep insight into essential problems. I once asked "What sort of philosophy are you studying?" He said it was epistemology. I asked, "What is it?" He explained various things, but I have forgotten them. It must be that a person so intently conscientious could not be satisfied with the limitations of logic and had to go beyond them.

A conference was held in 1966 commemorating Gödel's sixtieth birthday. I heard a story from Kreisel concerning it. An organizer asked him "Whom would you like to invite as guest speakers?" Gödel mentioned several logicians, but, when asked if he would suggest any philosophers, he answered "There are none, because there have been no noteworthy philosophers so far." I am not sure if it is a true story. Stories about Gödel tend to be exaggerated, and I believe this is one such example.

It appears that for Gödel his religion, view of life, philosophy and mathematics were unified. He believed in a beautiful harmony existing in the essence of the universe, and that was his religion, view of life and philosophy. He never ceased to believe in the victory of reason. That is that mankind will discover the beautiful essence of the universe. In that sense, mathematics is ranked at the top in the center of his view of life. (From now on, what I write is based on impressions I got through personal discussions with Gödel, and not what has been made public by him. It is quite possible that I misunderstood. I ask the reader to read it as *my interpretation* of "Gödel's opinions" rather than Gödel's opinions themselves.)

His philosophy even influenced his view of the axioms of set theory. For example, to Gödel, of the two fundamental principles of set theory, the principle of the ordinal numbers expresses the world of reason, and the principle of the power set expresses the world of chaos (or the world of reality). Since for him mathematics has to show the victory of reason, the principle of creating power sets must be controlled by the principle of creating ordinal numbers.—Considering this, he proposed a new axiom. (This axiom does not say "all sets are constructible." The idea of constructibility must have been devised according to a similar philosophical intention, but to Gödel the notion of constructibility is too narrow and hence it seemed to him too unnatural to be true that all sets are constructible.)

I cannot necessarily agree with him in such ideas though I discussed them with him repeatedly. Being an Oriental, I have rather been accustomed to dualism (the principles of "Yin and Yang"). Since set theory is composed of two completely different principles, the principle of creating ordinal numbers and the principle of creating power sets, would it not be more natural if we assumed that these two are entirely independent and different principles? To me, it is not an absolute necessity that the world of ordinal numbers control the world of power sets for the victory of reason to be achieved. Could we not call it a victory of reason if we discover a beautiful law in the chaos of power sets? Besides, being an Oriental, I can imagine a quiet nothingness in the world of power sets and am fascinated with that idea, rather than conceiving of it as chaos. This was my opposing argument. (I have proposed an axiom myself arising from this opinion.) Professor Gödel gave a wry smile saying that the Oriental philosophy was not nice if it was so pessimistic. (I presented the opposing position that western philosophy was too optimistic. Of course, there was no end to our argument.)

Another point on which we often argued was why he thought the cardinality of the continuum was small. His opinion on this was not so strong as the one above, but his assertion was that if there is a beautiful scale in the world of real numbers, controlled by reason, then such a conclusion necessarily followed. (However, he was not completely certain if there is such a beautiful scale in the world of real numbers.) I could not understand this notion thoroughly, while I was able to understand the preceding theory even though I disagreed with it. (I obtain somewhat different results when I examine the "scale." Furthermore, I simply could not grasp a "beautiful scale.") This was a regret I had when I left the Institute.

In recent years there have been few logicians in Princeton. So, to me Princeton means the place where Gödel lived. The Institute is the place where Gödel's office was located. When I recall Princeton with my eyes closed, I can see that beautiful office of his, the surrounding pond and woods, his gentle face and our mathematical discussions which I miss.

4.2   A. A. Fraenkel and Paul Bernays

4.2.1   *Fraenkel's axiom of replacement*

Both Fraenkel and Bernays contributed a great deal to the process of forming the present system of axiomatic set theory. As you may know, in naive set theory, which had been initiated by Georg Cantor, when one mentioned a set x, the scope of what x expressed was vague. Namely, while it had to be clearly decided, for another set y, whether "y is an element of x" or "y is not an element of x," it was not always possible as the scope of sets, or the definition of a set, was ambiguous. A set was considered merely as a collection of objects. It was assumed to be determined if y was an element (called also a member) of a set x, and if that was so, then one would write $y \in x$. In such a vague circumstance, given $P(y)$, a property of y, it seems appropriate to define a set a as the collection of all sets x satisfying the property $P(x)$. Namely, if we express this set a by $a = \{x|P(x)\}$, then whether x is an element of a can be clearly decided according to whether $P(x)$ is correct.

How revolutionary such an idea was and how it transformed mathematics creatively is evident from the history of mathematics. This idea, however, contains various contradictions. As for the famous Russell paradox, if one defines $a = \{x | x \notin x\}$, then one obtains

$$a \in a \rightleftharpoons a \in \{x | x \notin x\} \rightleftharpoons a \notin a,$$

which yields a contradiction. Where does this contradiction come from? What should we do about it? The first well reasoned solutions to this problem were given by Whitehead-Russell and by Zermelo.

These two solutions are essentially similar, although they assume different forms. We proceed as follows. Russell's paradox derives from the fact that the definition of a set is not clearly given. That is, when a set x is mentioned, the scope of what x expresses is not clearly defined. Therefore, let us consider something more specific when we consider a set. Namely, from now on, let us always assume that there is a domain D which already has been determined, and when we refer to sets, we will consider only subsets of D. In this way, the meaning of $\{x | P(x)\}$ is $\{x | x \in D \& P(x)\}$, and as for $a = \{x | x \in D \& P(x)\}$, we can use

$$x \in a \rightleftharpoons x \in D \& P(x)$$

at will.

Such an approach is practical, and plays a role similar to naive set theory[16] as far as those sets which are useful in concrete mathematics, such as the set of natural numbers and the set of real numbers, are concerned. At the same time, this way of thinking has the creative character of naive set theory, and leads us to a vast new world. Namely, let us start with a domain D to develop our consideration. (For example, let D be an empty domain, the domain of all natural numbers or the domain of all real numbers.) We can study subsets of D according to the principles above. We can therefore regard the totality of subsets of D as a clear domain. Let us denote this new domain by $\mathcal{P}(D)$, and call it the power set of D. That is to say, the elements of $\mathcal{P}(D)$ are the subsets of D. By applying Cantor's diagonal method, we can see that $\mathcal{P}(D)$ is far bigger than D and is a new world. Then, it must be that $\mathcal{P}(\mathcal{P}(D))$ is a new world even bigger than $\mathcal{P}(D)$.

What will happen if we continue to create ever larger worlds in this way? We will henceforth not use words like domain or world each time, but will consistently use the word set. Now, let D be D_0, the power set of D be D_1, the power set of D_1 be D_2, and similarly create $D_3, D_4, \cdots$. Then we can obtain bigger sets. When D_n's have been created for all natural numbers n, one can consider an object by joining all U_n's, that is the union

---

[16]The name "naive set theory" stands for classical set theory before it was axiomatized.

$\cup_n D_n$, and this must be a set larger yet than any of D_n. Denote this, for instance, by D_ω, and further continue to create $D_{\omega+1}$ as the power set of D_ω, $D_{\omega+2}$ as the power set of $D_{\omega+1}$, $\cdots$. Then we will obtain still larger sets. The universe of set theory is obtained by continuing this operation limitlessly, that is, without limit.

I have written "limitlessly" as one word. How then can I clarify "limitlessly?" It was Fraenkel who first presented a beautiful, though somewhat incomplete, solution to this difficult problem. Fraenkel's partial solution can be intuitively explained as follows.

Axiom of replacement   Let f be a function from a set D. Then, $\{f(x)|x \in D\}$ is a set.

This axiom satisfies what we mean by the expression "limitlessly" in the sense of first degree approximation. Let us consider, for example, the situation where we have created $D_0, D_1, D_2, \cdots, D_n$. Then a new set $D_{n+1} = \{f(x)|x \in D_n\}$ can be created by the axiom of replacement. We can thus create $D_0, D_1, D_2, \cdots, D_n, D_{n+1}, \cdots$ limitlessly. The problem is whether our limitless process has been completed with this. Let us denote the set of all natural numbers by N, and define $f(n) = D_n$. Then f is a function from N. Fraenkel's replacement axiom indicates that $D_0, D_1, D_2, \cdots, D_n, \cdots\cdots$ is not at all the totality of our limitless creation and that a D_ω, which is bigger than any of D_n, must be created. The situation is the same when we will have further created $D_\omega, D_{\omega+1}, \cdots, D_{\omega+n}, \cdots\cdots$. Namely, it turns out that we will have to further proceed to create a set which is bigger than any of $D_{\omega+n}$.

This way, our intention to create "limitlessly" can be satisfied largely by requiring a single axiom, the axiom of replacement. The history of mathematics to date has proved that, in fact in almost all fields of modern mathematics, this first degree approximation of "limitlessly" suffices. However, from the standpoint of modern foundations of mathematics, it is well-known that this first degree approximation is insufficient, and various candidates for the second degree approximation have been proposed by many researchers. But then no approximation of the second degree which is as splendid as the one of the first degree, which has proved to be an extraordinary success, has been discovered. This tells us how great Fraenkel's achievement was.

4.2.2   The works of Bernays

Now, let me switch to the achievements of Bernays. The system of axiomatic set theory which was established by Zermelo and Fraenkel was
different from what is called ZF set theory at present and was not mathematically precise. It was von Neumann and Bernays who established the
present well-organized system of axiomatic set theory based on the preceding set theory which had been stated in a vague form. Needless to
say, during this process, they had to prove many useful theorems within
set theory. What axiomatic set theory is, is now clear. For this reason,
Bernays' work tends to be underestimated. However, it must be evaluated
in historical context.

4.2.3   Fraenkel and Bernays from my viewpoint

A. A. Fraenkel died on the 15th of October 1965 at the age of 74. He
had been a professor at the University of Marburg in Germany, but moved
to Israel and rendered great service to the development of the University
of Jerusalem as a professor there. Brilliant Israeli logicians such as A.
Robinson, M. Rabin and A. Levy all studied under Fraenkel. I met him
twice at Logic Conferences at Stanford in 1960 and in Jerusalem in 1964.
He was a small old man with long beard, just as I had seen in a photo.
At Stanford, he chaired Robinson's invited lecture, and stated that he was
proud of A. Robinson being his disciple. In Jerusalem, he was very worried
about Bernays' absence due to illness, and implored the participants of
the Congress from the dais:"Let me ask those from Switzerland, to tell
Bernays upon your returning there that I wish him a quick recovery." His
initials A.A. are an abbreviation of Abraham Adolf, but I heard that it used
to denote Adolf Abraham and had been changed to Abraham Adolf since
Hitler. It is said that he was a close friend of the first Israeli President, and
so I imagine he was involved in the founding of the country. When I met
him, he was quite amiable, and said to me:"I have heard much about your
work from various people."

   I visited at the Institute in Princeton during 1959-60 for less than a
year, in the same period that Bernays was there. Schütte and Feferman
were also there. Rosser, Putnam and Smullyan were at the University.
Davis often visited us from elsewhere, and logic seminars in Princeton were
richly productive. I was sometimes perplexed when Bernays began to talk
in English and switched to German without noticing. In spite of his age,

he had an amazingly clear mind and awareness of the times.

Earlier I have remarked only on Bernays' work in set theory, as it was a continuation of Fraenkel's work. He served as Hilbert's assistant for a long time, and studied proof-theory diligently. Proof-theory in the period between Hilbert and Gentzen should be regarded as being represented by him. He was a small person, and everybody who met him spoke of him as gentle. Music was his hobby, and I heard that a Swiss orchestra had played one of his compositions. At a party at the Morse' residence in 1959, Bernays played Beethoven's "Pathetic" and one of my daughters played Bach. His piano performance was as sweet as his character. It is a pleasant memory. We discussed set theory, consistency problems, and various other subjects. Whenever I think of him, I recall strangely a bright spring day in Princeton with flowers in full bloom.

When I wrote in an article "Gödel set theory" about what is now called "Bernays-Gödel (BG) set theory," Bernays commented in person: "I do not mind when it is called Gödel-Bernays set theory but I feel resentment in my heart if it is called Gödel set theory. I defined the system, and Gödel only used it." I have certainly been determined to call it BG set theory since then. Even expressing such feelings, he maintained a calm and warm atmosphere, and I recall it fondly.

Bernays is a cousin of the famous psychologist Freud. Even now he is invited to many of the meetings and presents interesting lectures (although he is mostly absent and only submits papers.) The last time I met him was in 1966 at the congress of logicians in Amsterdam. He turned up only for the talks by Schütte and myself, and listened to me attentively. It was he who most earnestly asked questions after my talk. While I was pleased, I could not help recognizing that he had lost vigor and somehow felt that it would be my last chance to meet him.

4.3 Paul Erdös

4.3.1 *Extending Ramsey's theorem*

The mathematics of Paul Erdös covers a wide area, for example he is also noted for his contributions in number theory. Here I would like to write about his work in set theory. Since his work is an extension of Ramsey's work, I will first explain Ramsey's theorem.

Let N be the set of all natural numbers, and let r be a natural number. Define $[N]^r$ to be the set of all sequences $n_1, \cdots, n_r$ from N such that $n_1 < \cdots < n_r$. Let f denote a function from $[N]^r$ to $\{0, 1\}$. Namely, $f(n_1, \cdots, n_r)$ denotes either 0 or 1 when $n_1 < \cdots < n_r$. Under these conditions, there exists an infinite subset of N, say X, such that, given two arbitrary sequences from X, say $n_1, \cdots, n_r$ and $n_1' \cdots, n_r'$, where $n_1 < \cdots < n_r$ and $n_1' < \cdots < n_r'$, the equation

$$f(n_1, \cdots, n_r) = f(n_1' \cdots, n_r')$$

holds.

For $r = 1$ this theorem is equivalent to the usual "pigeonhole principle" but for $r = 2$ it is not so simple as it looks. (An elegant solution wanted!)

Erdös extended this theorem to an arbitrary infinite set instead of N.

Let me state one of his typical theorems, by assuming the generalized continuum hypothesis for simplicity:

$$\aleph_{\alpha+r} \to (\aleph_{\alpha+1})^r$$

for a natural number r. This expresses the following. Let us denote the totality of sequences $\beta_1, \cdots, \beta_r$ such that $\beta_1 < \cdots < \beta_r < \aleph_{\alpha+r}$ by $[\aleph_{\alpha+r}]^r$, and let f be a function from $[\aleph_{\alpha+r}]^r$ to $\{0, 1\}$. Then there is a subset of $\aleph_{\alpha+r}$, say X, satisfying that $\overline{\overline{X}} = \aleph_{\alpha+1}$ and, given two arbitrary sequences from $[\aleph_{\alpha+r}]^r$, say $\beta_1, \cdots, \beta_r$ and $\beta_1', \cdots, \beta_r'$, such that $\beta_1 < \cdots < \beta_r$ and $\beta_1' < \cdots < \beta_r'$,

$$f(\beta_1, \cdots, \beta_r) = f(\beta_1', \cdots, \beta_r').$$

It is somewhat difficult to explain the meaning of this theorem. Would it give a general idea if I say that it is connected to the extension of permutations and combinations to infinite sets? Such a way of thinking is useful in modern foundations of mathematics. For example, suppose a proposition $\phi(x_1, \cdots, x_r)$ on $\aleph_{\alpha+r}$ is given, and its truth value is independent of the order of $x_1, \cdots, x_r$. Namely, it is assumed that, if $x_1', \cdots, x_r'$ is a permutation of $x_1, \cdots, x_r$, then $\phi(x_1, \cdots, x_r) \leftrightarrow \phi(x_1', \cdots, x_r')$ holds. (For an arbitrary set, its elements can be assumed to be well-ordered.) Now, define a function $f(\beta_1, \cdots, \beta_r)$ by assigning the value 1 to it when $\phi(\beta_1, \cdots, \beta_r)$ is true, and the value 0 when $\phi(\beta_1, \cdots, \beta_r)$ is false. If we apply the above theorem of Erdös and obtain an X satisfying $\overline{\overline{X}} = \aleph_{\alpha+1}$, then we will know that $\phi(x_1, \cdots, x_r)$ is constantly true or constantly false on X. This idea is currently popular in set theory, and consequently old work by Erdös is in the limelight.

4.3.2 *My age is two billion years!*

I do not know the age of this Hungarian born old man. When I once asked him his age, he replied: "My age is about two billion years. For, when I was seventeen years old, it was said that the age of the earth was two billion years. But now the age of the earth is said to be four billion years. So, my age is approximately two billion years." Someone else then asked: "Please then tell us about reptiles and glaciers," but he evaded the question by answering: "The old man remembers olden times well, but has forgotten all about recent events." He has an extraordinary interest in the fact that, while mathematics is prospering in Japan, the income of the mathematicians there is much too low. He loves the story that, at the International Symposium on Algebraic Number Theory held in Japan [17], foreign participants presented an appeal to the Japanese Government to improve the working condition of the Japanese mathematicians. When I met him later, he immediately asked me "How is it? Has your income been raised?" When I answered "Not at all," he smiled wryly.

The problems in the range of his interest mostly have elegant solutions if the generalized continuum hypothesis is assumed, while they are unmanageable without it, and so he is very fond of the generalized continuum hypothesis. As Gödel proved its consistency (relative to BG set theory), Erdös refers to Gödel as "Pope, Pope." When later Paul Cohen proved its independence, he assigned him the honorific "Highest Cohen" (pronouncing it either in Latin or Hebrew). ("Cohen" is the Hebrew word for priest.)

His lectures are intuitive and interesting, but he switches from mathematics to jokes and from jokes to mathematics so rapidly that sometimes I get lost after a while. He himself seems to wonder "What in the world am I talking about?" When he proposes an open problem during a lecture, he offers a prize by saying "To a person who solves this $\cdots$." For example, to the problem

$$\omega^\omega \to (\omega^\omega, 3)^2$$

a prize of 250 dollars is offered. (In order to decrease the number of competitors, I will not explain the meaning of the problem. Please look up

---

[17]1955

his paper.) Once, when he proposed a problem with a prize, a Swiss man Specker solved it in one day and received the prize. (I do not remember whether it was 10 dollars or 100 dollars.)

At one time he visited the University of Illinois, remaining for less than a year, and attended a conference on set theory in Los Angeles in 1967. I attended his lectures on many occasions. His ideas were always intuitive and geometric, and so quite interesting. According to him, he is not suited for logic. At the conference in Los Angeles, I gave a talk which denied the continuum hypothesis, and he was disturbed by it. He kept asking for just the two of us to discuss it. Each time I had to turn down his invitation as I had another engagement. Even now I feel sorry about that.

Erdös knows just one Japanese sentence for some reason I do not know. He says "Boku wa idai na suugakusha desu" and smiles. He said to me "This means that I am a great mathematician. What should I say to mean that I was a great mathematician?" He restated then "Boku wa idaina suugakusha deshita" and smiled again.

4.4   Alfred Tarski

4.4.1   *The theory of truth values and model theory*

Tarski has a large body of work. Much of it involves interesting and clever applications of techniques of classical set theory. Here I will omit all of them and state the following two results. One is the theory of truth values and another is model theory. His theory of truth values was published in 1935. Many of his arguments have philosophical aspects as he is a distinguished philosopher, but I will not mention them here. In his article, he speculates on what truth and falsity are.

In our usual practice, when we claim $1 < x$, for example, x (or, x and 1) is (are) a subject(s) and $1 <$ or $<$ is a predicate. In most cases, what can be the object of thoughts or what can be the subject is determined, and it is always expressed by a variable such as x. In such a case, the predicate $<$ or a proposition is never regarded as the object of thoughts. This is necessarily the case when a logical system is fixed, and then what the variable x represents is never a proposition. Tarski investigated a logical system in which one can treat propositions of a given logical system as variables.

In order to do this, to state the conclusion, second order logic suffices if the original logical system is of first order, and third order logic suffices if the original one is of second order. He simultaneously proved that, if a logical system can express its propositions as variables within itself (as long as it satisfies some standard conditions), then that system is inconsistent. This serves as a proof of the fact that, given a logical system, another logical system in which it can be judged whether any proposition of the given system is true or not is necessarily stronger than the given system. This fact has a close relationship with Gödel's Incompleteness Theorem. The above mentioned article was published after Gödel's, and for that reason there was initially a criticism that it was an imitation of Gödel. In fact, he had originally written the article in Polish and the above article was its translation. [18]

This work is certainly quite significant, but after all, what should be regarded as his most influential results are in model theory. When we study group theory, we first learn a collection of axioms. We call a mathematical object which satisfies these axioms a group. In general, a collection of axioms (or propositions) forms a mathematical theory, and a mathematical object which satisfies the theory is called a model of that theory. In logic before Tarski, the major concern was the mathematical theory. However, in working mathematics, the emphasis is put on the individual mathematical object, that is, the model. Tarski therefore proposed to study model theory extensively in logic also.

This school was restricted to Berkeley until around 1960, and until then Tarski's reputation was not as great as later on. Subsequently, model theory made rapid progress, and at present it is a major subject. This appears to show that Tarski's basic idea was correct. Tarski's greatest contribution must be that he proposed model theory and taught many outstanding researchers, although Tarski himself may not have been completely satisfied. We might say that, by virtue of the rise of the Berkeley School, logic in America was reanimated after the decline of the earlier Princeton School.

4.4.2 *The motive power to promote logic*

Apparently the way Tarski educated his disciples was demanding. I have heard many incidents of brilliant students who gave up due to Tarski's demands. He had the reputation of being a difficult dissertation advisor.

---

[18] Author's Note: According to Susumu Hayashi, in the original article in Polish, it was not proved that the logical system in which the truth value can be defined of a given logical system is stronger than the original one. My impression as above comes from what Gödel told me and from Mostowski's book.

For example, he suggested a problem to a student saying its solution would constitute a Ph.D. dissertation. The student solved it in three months, only to be told by Tarski that he could not grant a degree for a problem that was solvable just in three months.

At conferences, he would glare at the participants. He had a forbidding manner which kept people other than his students at a distance.

It is interesting that a person with Tarski's personality was in a way the engine which motivated logic in America. To date there are only two logicians who have belonged to the American National Academy of Sciences, Gödel and Tarski. (This tells us that until recently there was a strong prejudice in America against logic.) [19] After Paul Cohen's revolutionary work, many young scholars in logic have emerged, and I have not heard much about Tarski recently.

I lived half a year in Berkeley and I also met him at various meetings, but as I did not talk much with him, I cannot recall many personal incidents.

One incident I do remember is his speech at the dinner party of the Berkeley Symposium in 1963. As he stood up and gave an address as the director of the symposium, he spoke tearfully about his memories of the many logicians who had been victimized by the Nazis. He is of small stature like the other logicians I have described so far. On one occasion, I asked him "What do you think of intuitionism?" His reply to this question was "Philosophically weak, and mathematically ugly." This clear-cut answer was characteristic of him and it made a strong impression on me.

4.5   A. Heyting

Heyting is one of Brouwer's disciples and a carrier of intuitionism's banner.

Let me first explain Brouwer's intuitionism. We believe in the excluded middle "A or not A." We might say that it is because we assume the existence of an absolute standpoint like that of God. (Logic is considered from such a standpoint.) Namely, for the Absolute it is known "either A holds, or A does not hold" and so it is obvious "which is correct."

However Brouwer's intuitionism regards such a logic (or mathematics) as too transcendental, and proposes to consider logic from a more human standpoint. That is, if we understand that claiming A does not mean it is true from the standpoint of the Absolute but it means it can be proved from the human standpoint, then the excluded middle must be understood

---

[19]The date of this statement is not clear.

to claim that either A or the negation of A can be proved. This is not at all obvious. It was Brouwer who created an entirely new logic and mathematics by applying such an interpretation not only to logic but also to natural numbers, real numbers and in other areas. Such logic and mathematics are called intuitionistic.

Heyting is a disciple of Brouwer, and he is the one who has passed on intuitionism to the present. Brouwer's intuitionism had a rather personal flavor, but we might say that intuitionism has become objective through Heyting. Heyting's accomplishments are that he first formalized intuitionism and made it easier for the outside world to appreciate its outlook, and, more essential yet, that he clearly formed the foundations of modern intuitionism by emphasizing the notion of "constructive." Heyting demonstrated that all of Brouwer's ideas could be rationalized by interpreting, for example, $\forall x \exists y A(x, y)$ (for all x there exists y which satisfies $A(x,y)$) as follows. $\forall x \exists y A(x, y)$ holds if a constructive method is given so that, for any x, it constructs a y such that this y satisfies $A(x, y)$. This work is important as it has clarified that the essence of intuitionism is "construction."

At present, fresh trends in intuitionism are manifested by the activities of Kleene, Kreisel, Myhill, Troelstra and others. We might say that they are considering the areas of intuitionism which Brouwer could not sufficiently develop, by extending Heyting's ideas. On the other hand, Heyting himself, having completed an interpretation of Brouwer's ideas by "construction" and its formalization, seems to have gone in the direction of doing mathematics, such as topology, integration theory and the theory of Hilbert spaces in an intuitionistic setting.

Heyting was a professor at the University in Amsterdam until 1968 and trained many researchers in intuitionism, but some have criticized him as being too traditional, and suggested as a consequence that his disciples have not succeeded as expected. In Buffalo last summer[20], there was a symposium on intuitionism and proof theory, and Heyting showed his pride in his student Troelstra, who was becoming a new star in the area. I have only met Heyting at the conferences in Stanford, Jerusalem, Amsterdam and Buffalo, and so am not well-acquainted with him personally, but he is a refined and courteous old gentlemen. His wife is much younger than he, and both she and his daughter are of large build, in contrast with Heyting's slightness. Mrs. Heyting seems to be vigorous and outspoken. She commented on a portrait of Brouwer: 'He looks like Mr. Hyde in "Dr.

---

[20]1968

Jekyl and Mr. Hyde",' and that of Hilbert: 'Ah, he resembles Lenin.'

During the congress in Amsterdam last year, Heyting was quite busy serving as the chairperson and appeared tired. When I met him last summer in Buffalo, he had retired from the University of Amsterdam (at the age of seventy), and he looked sad. He approached me and asked about recent developments in proof theory. Then, having told him a bit, I in return said "Congratulations!" for recent remarkable developments in intuitionism. He smiled very happily.

4.6   Alonzo Church

Before I write about Church's works, I would like to mention that Church and H. Curry were the pioneers of mathematical logic in America. Both of them studied logic with Hilbert in Göttingen and brought it back to America. Of them, Church was able to take the advantage of the location of Princeton University, and trained many brilliant logicians such as Rosser and Kleene, thus developing the excellent Princeton School. (Of course, it is obvious that Gödel was an influence as well.)

Now, Church's work must be above all "Church's thesis." Church introduced the "λ-operator." In the current terminology, this corresponds to "recursive functions." In this sense, we should regard Church as one of the creators of the theory of recursive functions.

What is important about the recursive function is that its mathematical definition is accurately given. "Church's thesis" asserts that recursive functions and "computable functions" coincide.[21] Of course, the notion of "computable" is not defined, and so there is no way of proving this thesis. However, we have a vague idea of computability, and it would be indeed marvellous if it could be mathematically defined. Later this thesis became a generally acknowledged principle after A. Turing (another disciple of Church) had defined the "Turing machine" and proved that the functions computable by Turing machines and recursive functions coincide. Nowadays, negative solutions of the "word problem" and other problems are given based on Church's thesis.

It is quite a rare thing that a notion such as "computability" could be defined so easily. Gödel was impressed with the absolute nature of recursive functions which does not depend on any systems, and observed that this was "by some miracle". We might claim that, the theory of recursive functions

---

[21] Here functions are confined to number theoretic functions.

came to us through the philosophical motivation of Church's thesis, to which it owed its robust growth.

Church is of majestic build. He is a man of few words and sunny disposition. His speech is so soft that it is difficult to become friendly with him. His wife is also of large build, is cheerful, moves quickly and speaks merrily with a clear voice.

When we were invited to their house, Church was seated heavily on a large chair and hardly talked or moved. Mrs. Church alone offered subjects of conversation, moved about busily and took care of guests. (This is rare in America.)

Mrs. Church is originally from Poland. She is a cheerful and friendly person, and invited us to her house often. In the beginning, she addressed us properly as Takeuti, but eventually she changed it to Takeyuki and finally to Sukiyaki. Church himself heard this and smiled, but said nothing. So, I eventually corrected her.

Church was a professor at Princeton University for a long time, but moved to UCLA (University of California at Los Angeles) in 1967. I visited Princeton in 1966. As it was Church's last year in Princeton, he had no students and Church himself was more interested in the history of logic than logic itself. More precisely, he devoted himself to conscientiously editing reviews of articles in logic in order to have them published in JSL (Journal of Symbolic Logic). I gather the reason why he moved from Princeton to Los Angeles was that in Princeton he could not get sufficient support for the work of reviewing for JSL. As I heard at UCLA, he was concentrating on editing reviews of articles, and hence did not attend logic seminars.

Incidentally, his daughter is Mrs. Addison, another logician. Addison is a student of Kleene, and so is a grand-student of Church. After Church had left Princeton, an outstanding young logician Kochen succeeded him, but it cannot be denied that the long standing traditional Princeton preeminence in logic has been interrupted.

4.7   S. C. Kleene

Kleene is the person who essentially created the theory of recursive functions. It is no exaggeration to say that the theory of recursive functions in a narrow sense was completed almost by Kleene alone. It is mainly due to his effort that today recursive functions are common knowledge and serve as a useful tool in logic. After his research on recursive functions, Kleene played an active role in their extension and application, but now for some time has been devoting himself to intuitionism.

As I stated before, the central notion in intuitionism is "construction," and the central idea of recursive functions is also "construction." In that sense, recursive functions and intuitionism are related but Brouwer's intuitionism had been created before the recursive function was born, and hence the notion of the recrusive function is not used at all in it. Today, as the centrality of the recursive functions has been established, it is significant to revisit intuitionism with recursive functions in mind. It appears that Kleene's project was initiated by such a notion, and that it, together with Heytings's work mentioned earlier, forms a foundation for the latest developments in intuitionism.

Kleene is a giant. He is slim and his voice is loud, so that one gets an impression of him as an electric light pole with a loudspeaker. His given names are Stephen Cole, but he does not seem to like these names, and always abbreviates them to S. C. I have never heard anyone call him Steve. It is said that his lectures have a matter-of-fact tone. I have also heard that he writes on the blackboard from his notes, never looking towards the students, so there is no chance to ask questions. However, his talks at conferences are not like that.

When I stayed at his house in 1960, he appeared to be pressed by work from morning till night. He worked on his current research, took care of his written articles for publication, did chores for the department, and so on and so forth. (Very often he serves as the chair of the department and the president of an academic society.) I had the impression that it all made him seem like a machine. He does not drink alcohol, does not smoke, and does not drink coffee (He mostly takes milk.). He said that, although he had been fond of playing chess, he soon quit it as it became an obstacle to his studies. I honestly wondered how a person could work like that all the time. Since his wife had a job, he cooked eggs and toast for all of us (My whole family stayed at his place.). When he asked us :"How hard should the eggs be? How dark shall I make the toast?", I felt the United States, where such a great scholar served us like that, was striking rather than admirable.

In the summer of 1960, I gave a talk in Kleene's seminar on a theory of recursive functions over ordinal numbers. He was very pleased with it and promised to study it himself soon. Later at conferences in Jerusalem and Amsterdam, he always sat at a front seat and listened to my talks attentively with a big smile. I felt glad as I thought he was interested in my theory. However, in fact it was not so. He was one of few mathematicians to attend all the talks at any conference. I learned that fact when I gave a series of lectures for two weeks at the conference in Buffalo. When he was really short of time, he would start proofreading his books and articles. Toward the end of my group of lectures, he devoted himself to proofreading from the beginning. I felt a bit mischievous, and inserted in a loud voice something like "Just as Professor Kleene did with recursive functions" in my lecture, but he remained still and focused on proofreading.

4.8 Georg Kreisel

Here I would like to make an excuse. As I wrote at the start, the reason I began to write this essay is that I read about my good friend Kreisel in "The Double Helix"[22] and that fact stimulated my intention to write memoirs of my association with close friends at my convenience.

However, as I first began to write about the esteemed Professor Gödel, the effort became quite serious and the following question occurred to me. What will happen if I continue to write in this way? There will be close acquaintances of my generation such as Mostowski, Schütte, Robinson, Vaught, Scott, Feferman, ······, then a star Cohen, and after that promising people like Solovay, Silver, Martin, ······. There are too many. Furthermore, it is difficult to write about their recent works in this mode, and, as for young people, I should put emphasis on the future rather than present circumstances. Having made this judgement, I decided not to undertake my original purpose of writing "memoirs of my association with close friends," but rather put it aside for another occasion.

Another excuse I would like to make is that I have not taken up anyone whom I have not met, however deeply I respect that person. For instance, I revere Gentzen as much as Gödel and I have learned about him through Schütte, Bernays and others, but I have not included him here. The same with Brouwer and Turing. I wished to maintain the authenticity of my observations.

Because Kreisel caused me to start these articles, I will treat him specially, although he is of my generation, and I will close my personal observations with him.

Kreisel's major achievements can be classified into the areas of proof

---

[22]Published in Penguin Books with new Introduction 1999

theory and intuitionism. The virtue of his work in intuitionism is that it drew attention to this area and that it opened a new kind of intuitionism. I will not evaluate this new kind of intuitionism, as these are issues for the future. It seems that Kreisel's own attempt is to create a rather strong type of intuitionism. His most remarkable achievement in proof theory is that he proposed various problems, which stimulated Feferman, Tait, Howard, Friedman and others, and made the area prosper.

Kreisel is a bright person. There are numerous intelligent observations in his articles. He is philosophically inclined, and hence always speculates on the philosophical meaning of a mathematical problem and as a result raises ingenious new questions.

Before I comment on his person etc. in detail, let me cite the lines concerning Kreisel from "The Double Helix"[23] as I mentioned earlier.

> By then we had been joined by Francis's close friend, the logician George Kreisel, whose unwashed appearance and idiom did not fit into my picture of the English philosopher. Francis greeted his arrival with great gusto, and the sound of Francis's laughter and Kreisel's Austrian accent dominated the spiffy atmosphere of the restaurant along High Street at which Kreisel had directed us to meet him. For a while Kreisel held forth on a way to make a financial killing by shifting money between the politically divided parts of Europe. ······ and the conversation for a short time reverted to the casual banter of the intellectual middle class. This sort of small talk, however, was not Kreisel's meat, and so······

and much later it goes as follows.[24]

> This time the correct equations fell out, partly thanks to the help of Kreisel, who had come over to Cmabridge to spend a weekend with Francis.

I heard that "The Double Helix" is written in such a forthright manner that many people felt injured and filed lawsuits. I also heard that they asked Kreisel to join in suits, but he refused.

As is written in "The Double Helix" above, Kreisel is said to be very knowledgeable about food, and to amaze others with his knowledge of wine.

---

[23] pp.69-70
[24] p.116

He is a rare kind of a mathematician, as he is acquainted with many celebrities such as wealthy people and actresses. As for me, he is, fortunately or unfortunately, sometimes a rival and sometimes an ally in proof theory. Once a discussion starts via letters, he sends letters to me almost every day. (Of course I too write letters almost every day.) In one extreme case, I received three letters in one day.

He was born in Austria, but went to England with his father at the age of seventeen, parting from his mother. I have heard that, after the Second World War, he looked forward to visiting his mother in Austria.[25]

He has been a professor at Stanford University for a long time, but for a while he taught at the University of Paris and Stanford University every other year. Now he does not hold a position at the University of Paris, but every year he spends much time in Europe. (He says he earns in America and enjoys in Europe.)

Now (the spring of 1969), for example, he is in Europe. He commonly visits Princeton on his way back from Europe to talk to Gödel. At various conferences and articles, he quite often voices some opinions and I am not certain if they are his own or those of Gödel's. He claims he is the new god of intuitionism succeeding Brouwer, and speaks and responds with an air of importance. He is quite a performer, and I must say that his ability to achieve the status of a celebrity is admirable.

It is often said that it is impossible to read his articles, and that it is not clear what is written in them, but his writings seem to have improved recently. My articles are also said to be difficult to understand. Once I spoke with Gödel about it, and when I told him "Kreisel's articles are difficult to understand because his English is too skillful, while my articles are difficult to understand because my English is too poor," Gödel was pleased and responded: "That may be so." I in fact intended to make an ironical comment on Kreisel's writings being startling in appearance and deliberately twisted. I heard that, upon hearing this, Kreisel made a comment: "This time, Takeuti's English was too good!"

His insomnia is famous. When he is invited to a place, he examines the hotels in the city one by one, and inconveniences his hosts by claiming "They are all no good." Satisfying him is indeed difficult, as a glimmer of light or a faint noise disturbs him. An acquaintance of his was awakened at two a.m. by Kreisel and asked to put him up for the night, because he could not sleep at the hotel under any circumstances. They examined all

---

[25] Author's Note: I heard that it had been realized only just before her death.

the rooms in the house and decided that the kitchen was most suitable. Then they worked hard, affixing a black curtain which Kreisel had brought along, carrying in a bed, and carrying out a refrigerator.

Once I start writing about Kreisel, there is no end to it, but many of the incidents are personal, and so it will be better to stop now.

Chapter 5

Set Theory and Related Topics

5.1 The meaning and significance of the axiom of determinateness

The greatest achievement in modern set theory during the past five years is the axiom of determinateness, which has been worked on mainly by Woodin, Martin and Steele. They were awarded the Karp Prize by the Association for Symbolic Logic.

The axiom of determinateness is quite a strange axiom. An ordinary axiom of set theory such as the axiom of replacement or the axiom of infinity is an axiom which directly refers to sets, but the axiom of determinateness does not appear to be an axiom of set theory. An ordinary game terminates within a finite time, while the axiom of determinateness refers to games which take infinite time to be played. The axiom of determinateness states that, in such a game, one of the players has a winning strategy. [1]

It seems that human beings love games, win or lose. It is said that in playing a game a special part of the brain operates, and I believe that games have a quality that fascinates people.

At any rate, if we assume the axiom of determinateness, many difficult problems can be solved as a consequence. It is extraordinarily special in that sense.

It is, therefore, entirely different in substance from previous approaches to the subject and hence it is an extremely difficult question how this axiom fits into set theory.

The existing axioms of set theory are, even if some new axioms are included, rarely directly useful to mathematics, but some of the consequences of the axiom of determinateness are broadly applicable to mathematics.

---

[1] A player wins if the chosen sequence belongs to a certain set of sequences.

For this reason, there arises the question how the axiom of determinateness is to be integrated as an axiom of set theory. To clarify, the axiom refers to infinite two-person games, in which each of two players alternately chooses a natural number. When infinitely many steps have passed, the whole process of the game is represented by an infinite sequence of natural numbers. In that sense, the axiom of determinateness expresses a property of infinite sequences of natural numbers. If we consider a two-person game in which each of two players chooses a natural number alternately and which halts within finitely many steps, then it can be shown that there is a winning strategy for such a game. As for an infinite game, on the other hand, one cannot decide about the infinite future, and this phenomenon appears to be related to the idea of randomness. On the other hand, since considering infinite sequences of natural numbers is equivalent to considering real numbers, the axiom of determinateness can be viewed as a new axiom stating a property of real numbers which was never previously thought of.

As the study of the axiom of determinateness has progressed, the following fact has become known. The axiom of determinateness ultimately determines a property of subsets of the natural numbers, that is, the power set of the set of natural numbers.

In truth, the axiom of determinateness itself is in fact not correct in the sense that it contradicts the axiom of choice. More explicitly, assuming the axiom of choice, one can construct a two-person game for which neither player has a winning strategy. If the axiom of choice is assumed, then all the candidates for the winning strategy can be ordered, and, using this fact, a counter-example can be constructed. The real situation is the following. It can be shown that there is a minimal model of set theory which contains the real numbers. The axiom of determinateness in the true sense claims that it holds on such a model.

The axiom of determinateness expresses a property of the real numbers or of the power set of the natural numbers which is entirely new and is of a form entirely different from existing axioms of set theory such as the axiom of replacement or the axiom of infinity. In that sense, it is an axiom which expresses a new and quite mysterious property of infinite sets, which was never considered in set theory before. This is the unanimous view of set theorists.

This introduces a highly important problem. For, in the progress of axiomatic set theory, such a problem has never been taken up before, and it is an axiom of set theory of a totally new kind.

Ultimately, the problem is whether it harmonizes with the axioms of existing set theory. Can the axiom of replacement in the traditional form or in its extended form justify a property of the real numbers or of the power set of the natural numbers which is completely different from the known properties? This problem is a grand challenge.

One after another, curious phenomena which could not have been derived in traditional set theory have been derived from the axiom of determinateness. For example, strong properties in projective set theory and in measure theory have been discovered. For set theorists who had worked with the old axioms such as the axiom of replacement, this has seemed to be a mysterious phenomenon.

As for projective set theory and measure theory, the consequences of the familiar axioms of set theory are very mild, while many extremely strong consequences of the axiom of determinateness have arisen. Therefore, the first reaction of set theorists was that the axiom of determinateness would be completely unrelated to and independent from existing set theory, and hence would be an axiom of a different kind.

However, as the study of the axiom of determinateness has progressed, the relationship between the axiom and existing set theory has begun to be brought out. For example, there is an axiom which claims the existence of a measurable cardinal. This is thought of as an extension of the axiom of replacement, and is believed to be true by contemporary set theorists. The existence of a measurable cardinal is not so powerful as the axiom of determinateness, but some of its consequences seem to be related.

It was believed that the axiom of determinateness is a shatteringly powerful axiom, so powerful that it was outside the realm of traditional set theory, and it could not be understood in the context of traditional set theory. In the last seven to ten years, however, things have been clarified and, after all, the axiom of determinateness can be derived from the axiom called the axiom of super-compactness, which had been considered in the framework of ordinary axiomatic set theory and which is quite an ordinary notion. This has settled the matter.

This result was quite unexpected. For, at first, these axioms had been thought unrelated, and then, even if they were related, it was thought that the axiom of determinateness would be much stronger. However, in the end, it turned out that the existence of a super-compact cardinal is much stronger than the axiom of determinateness; in fact the latter is only slightly stronger than the existence of a measurable cardinal.

This was surprising, and, a big victory for set theory, furthermore. There has been an idea, which was originally claimed by Gödel and others, that, if one added an axiom which is a strengthened version of the existence of a measurable cardinal to existing axiomatic set theory, then various mathematical problems might all be resolved. Theoretically, nobody would oppose such an idea, but, in reality, most set theorists felt it was a fairy tale and it would never really happen. But, it has been realized by virtue of the axiom of determinateness, which showed Gödel's idea valid.

In conclusion, we might say that modern set theory has made remarkable progress through Gödel and Cohen's work, and has been settled for the time being.

5.2 The future and Gödel's obsession

Now the problem is what to do next, and it is quite difficult to decide. There have been minor and straightforward developments, but there is no clear vision of major trends. Nevertheless, we can at least state one central problem, even if it does not help us predict anything.

There is a problem which Gödel considered as a life long concern. It is said that towards the end of his life, Gödel was obsessed with it as if he had lost his mind. It was Tarski who told us of this. There was an incident that proved it.

I have once mentioned that incident in "On Gödel's axiom" ("Sugaku Kisoron no Sekai" [2] : Nippon Hyoron Sha,LTD). However, nothing is written there about the reason why others think it is Gödel's obsession. The biggest issue with which Gödel was concerned was to present a new axiom on large cardinals such as the existence of a measurable cardinal as previously mentioned, in order to vanquish the problem of the continuum hypothesis.

Of course, in the usual sense, the continuum hypothesis cannot be proved or disproved in existing set theory, according to Gödel's and Cohen's theorems. Gödel was certainly well aware of this, so then in another way, Gödel wished to find a new axiom, especially an axiom for large cardinals, which might resolve the continuum hypothesis. Gödel concentrated on that idea.

There are two principles of set theory, one the operation of constructing power sets and one the operation of constructing ordinal numbers, thus making it possible to repeat the construction of power sets infinitely. When

---

[2]The world of the foundations of mathematics

you read this, you might think that, set theory can be understood clearly with these two principles, but in fact nothing is really clarified by them. The problem is therefore what a new axiom of set theory should be.

The reason why nothing can be understood from these two principles is that, given an infinite set, although one could talk about listing all its subsets and then reflecting on them all, in fact no human being can do it, and nothing can be grasped about what the power set is.

In that sense, the construction of all the subsets of an infinite set and the discussion of their properties must be called an armchair theory. That is exactly the reason why it is a big issue what the axioms of set theory ought to be, or perhaps what the principles of set theory ought to be.

If set theory were something within one's reach, then it would cause no problem and its axioms could easily be determined, but indeed such a thought is of no help. One can talk about it, but in reality an infinite set is not within a human being's grasp and nobody knows how an infinite set looks.

Therefore, although there are various properties crucial to the nature of set theory, the most essential among them concerns the nature of the power set. The simplest question then is what its cardinality can be. Whether the continuum hypothesis holds or not is, in a sense, the most basic problem in set theory and the sharpest question.

Set theory was invented by Cantor, and the image of the universe of Cantor's set theory may be referred to as "Cantor's paradise." On the other hand, Gödel cherished his image of the universe of basic set theory, that is, he believed that the axiom of the large cardinal would determine everything. Let us temporarily call it "Gödel's paradise." I claimed earlier that the axiom of determinateness yielded a big harvest in set theory. It can be regarded as a powerful realization of "Gödel's paradise" in the realm of thought.

For that reason, if there is to be a next step (not that anybody has started on it, nor that there is any clear movement), and if there is to be major progress, then it will be a clearer realization of "Gödel's paradise." That is, the determination of the status of the continuum hypothesis, which was Gödel's obsession throughout his life, will be resolved through the discovery of a new axiom on the large cardinal, just as I have mentioned.

Gödel worked painstakingly on the continuum hypothesis in his later years and examined various approaches. I am acquainted with them in detail, and I know that it is not a simple matter. Nevertheless, if this problem were to gain prominence through an effort to realize "Gödel's par-

adise," then it would become a really interesting problem, and I can claim
that it would be an important issue for set theory in the 21st century.

Once there was "Cantor's paradise," in which the foundations of mathematics originated and which is deeply related to proof theory, the area of
my research. What will be most interesting to me in connection with set
theory is, therefore, to what extent the realization of "Gödel's paradise"
progresses, or in more detail, if a large cardinal axiom will be proposed in
terms of which the problem of the continuum hypothesis is resolved.

The problem I have discussed so far is not merely a technical problem, but in a sense requires something like an encompassing view of the
world. It is certainly a difficult problem, but not just difficult. It cannot
be understood without a grand world view, a global way of thinking.

I have said, this is a problem belonging to the 21st century. It is an
exceedingly interesting but difficult problem. What will become of it? Can
we progress? Can "Gödel's paradise" be realized to a certain extent? My
opinion is that this is a great challenge.

For example, I believe that, supposing a resolution of Gödel's obsession
is achieved within the 21st century, many mathematical problems which
have remained unsolved so far will be solved in large numbers, not only
in the foundations of mathematics but also in mathematics in general. If
"Gödel's paradise" is indeed achieved, then by means of such a new axiom of set theory, many mathematical problems which would never have
been resolved without it will be solved. That is the paradise I hope for in
mathematics.

Actually, by applying the axiom of determinateness or the existence
of a super-compact cardinal, many problems which could not be resolved
without it have been solved. In that sense, the problem of the axiom of
determinateness signifies, as a realization of "Gödel's paradise," that set
theory has had great influence on working mathematics. That is what I
call "Gödel's paradise."

Thus, even if you do not know the foundations of mathematics or set
theory, once you have understood the meaning of the axiom of determinateness, you can design a game using entities from your field and derive
various theorems in your field from it. There is a real possibility of this
being a new mathematical methodology.

Whether an analyst or a number theorist, any mathematician will have
a powerful tool for making discoveries in his or her own field, once he/she
has understood the meaning of the axiom of determinateness. For example,
it is possible to define a game concerning analytic sets or measure, and then

to derive mathematical properties from the existence of its winning strategy from the axiom of determinateness. The axiom of determinateness is very close to all fields of working mathematics, and hence its applications have enormous potential.

At the moment, most mathematicians do not know about the axiom of determinateness, and that is why such a state has not been achieved, but my view is that if they become aware of it, it will be. In that sense, I think that set theorists should launch a campaign to make the mathematical community aware that there is such a wonderful axiom and that mathematicians may discover valuable theorems if they apply it.

5.3 Set theory and computer science

As for computers, let me start with their relation to the preceding account. In case of an infinite set, it is impossible to list all its subsets, and a human being cannot know them all. In fact, the situation is similar with a finite set.

In case of a finite set, it is of course desirable to list all its subsets. For, then, at least theoretically all problems regarding finite sets can be solved.

In practice, however, this is quite inefficient. So, in reality, it is impossible to solve all such problems with computers. Construction of a power set corresponds metaphorically to the exponential in the finite world, and hence almost all finite problems could be solved in the world of the exponentially large. Nonetheless, it is impractical to carry out such resolutions with computers, as it is too inefficient.

Now, manageable quantities can be estimated by polynomial functions, and not by exponential functions. The exponential function is, in principle, powerful enough to solve all the finite problems but is not practical, and so realistic methods of solving problems with computers involve using functions of polynomial type.

Accordingly, I will classify the problems which are decidable by programs of polynomial running time. The first class consists of the problems which can be decided by purely deterministic Turing machines within polynomial time. The totality of such problems is denoted by P. The next one consists of the problems which can be decided by non-deterministic Turing machines within polynomial time, and the totality of such problems is denoted by NP. We thus define in sequence more and more complex classes of problems which can be decided within polynomial time but by more and more complex machines. The most fundamental and difficult problem concerning this computational complexity is whether

$$P = NP$$

holds.

Now, let us return to set theory. As I observed previously, it is the power set of an infinite set that cannot be grasped by the human mind, and hence is a hard problem. For that reason, Gödel invented in his set theoretic image of the world the class L of all the sets which can be constructed in a clear manner. With this L, there is no hard problem inherent in the power set, and hence it is no exaggeration to say that all about L can be understood once the ordinals are well studied. I have in mind an analogy that P in the problem of computational complexity and L in set theory correspond to each other. Namely, consider the notion of relative constructibility as an extension of Gödel's L. This means the class of all the sets which are constructed with the operation that constructs L starting from a given set. L is a special case where the starting set is the empty set ϕ. NP is the class of problems which can be solved by first giving a random guess and then by applying the operation of P. So, if we make a correspondence

$$P \longleftrightarrow L$$

then the exact analogy

$$NP \leftrightarrow \quad \text{relative constructibility}$$

holds.

In this sense, there is a profound relation between a problem of computational complexity and a problem of set theory. The situation is as follows. Since it is impossible to investigate the power set or the exponential function thoroughly, one should start respectively with L or P, whose properties are well known, construct a hierarchy which reaches the power set or the exponential function respectively, and then study it. The two kinds of problems above are related in the sense that various notions arising in the process of so doing resemble one another. So, I think it can be claimed that the problems of set theory and the problems of computational complexity deal with the same fundamental issues occurring in different settings.

Now, the problem

$$\text{whether} \quad P = NP \quad \text{holds or not}$$

is, as has been said above, a fundamental and important problem. Not only that, I can also assert the following.

Consider Fermat's Last Theorem and the Riemann Hypothesis, which are not only important but also have some kind of magical power that has influenced the history of mathematics. Like these I believe it is certain that the problem of $P = NP$ has a great power which forces mathematics to progress. Computer science now comprehends a wide range of subjects, and it has various relationships with logic in the sense of the above.

It would be nice if I were acquainted better with computer science and could predict what will happen in the future and what are the biggest problems, but to my regret I am a logician and not a computer scientist. So, I cannot foresee that. I can at least say that this is one extraordinarily and decisively important problem.

5.4 Summing-up

Set theory, computer science and proof theory are respectively different on the surface, but in a sense they are close to one another. Although set theory and computer science are farthest apart, we can relate them with the following correspondences: the set versus computational complexity, the power set versus exponential function, L versus P, and relative constructibility versus NP.

As for computer science and proof theory, I am interested in computer science in relation to proof theory. Proof theory primarily concerns detailed studies of set theoretic problems, by investigating the inferences involved in them using refined and concrete methods so that they become clear as if they were laid in our hands. One cannot take a set itself in one's hand, but one can observe an inference in hand, and proof theory is a study of inferences. Inferring step by step and computing step by step amount to the same thing. In that sense, proof theory and computer science are directly related.

Although progress of various kinds occurs in the respective areas of these different fields, they all eventually flow into mathematics. Things which seem independent on the surface are related in some sense. That is why mathematics is interesting. So far, I have approached every subject in relation to proof theory, and it makes proof theory all the more interesting to me.

Chapter 6

From Hilbert to Gödel

The first research direction of my career was in the foundations of mathematics, especially Hilbert's Program to resolve the consistency problem.[1] The academic view then current was that this problem had been proved to be insoluble by Gödel's Incompleteness Theorem, hence no one took it seriously. I proposed my fundamental conjecture "The cut elimination theorem for higher order predicate logic" as a new project, and I began to do the research, feeling constantly alone and isolated. Gödel understood my research better than anyone else and was well disposed toward it. Gödel showed unusual interest in my work, and almost all our discussions centered around my approach to the consistency problem and set theory. I have no doubt about Gödel's strong interest in Hilbert's Program. It is also true that the reason why I came to America the second time to live permanently was my respect for Gödel. As can be seen from what I have said above, Hilbert and Gödel were major influences in my life.

6.1   Hilbert 1930

In 1930 the mantle of leadership in set theory passed from Hilbert to Gödel. This historic transition was a major event in the foundations of mathematics.

In 1930, Hilbert was sixty-eight years old. Sixty-eight was the retirement age in German universities. Hilbert was a great mathematician with deep insight. He always solved problems by treating them as parts of a deeper whole. Hilbert's magnetism drew many young geniuses to Göttingen. Göttingen, whose tradition extended back to Gauß, became

---

[1]Late 1940's, the post war time: See Appendix B.

truly the world center of mathematics because of Hilbert.

The disorder after the First World War was only beginning to subside, but in Göttingen a street was named "Hilbert Street" in order to honor this great mathematician.

Since 1920, Hilbert had concentrated on the foundations of mathematics. Hilbert was worried more than anything else that modern mathematics, whose value he believed in, might be damaged by the inconsistency of set theory. He could not tolerate the idea that Brouwer's intuitionism required the denial of the excluded middle and its consequences. Metaphorically, Hilbert's state of mind could be said to have been, "I cannot give up the paradise created by Cantor so easily."

6.2 Hilbert's Program

How should the foundations of mathematics be approached? To this problem, Hilbert proposed the following program, now called Hilbert's Program.

(1) Formalize mathematics and treat it as a formal system.
    For this reason, Hilbert's standpoint is called formalism.
(2) In this formalized system, a proposition is expressed with a string which is composed of some special symbols which stand for basic propositions, variables and logical notions respectively. Mathematical inferences and proofs are all described as concrete rules which apply to these strings, which are clusters of symbols as above. [2]
(3) Only the arguments which are applied to the formalized proofs that are concretely considered as above and which are simple and reliable can be admitted. Namely, the strings which are relevant here are concrete strings consisting of finitely many symbols. This standpoint is called the finite standpoint.
(4) Complete the foundational underpinnings of the formalized mathematical system by proving that no contradiction can be derived in this formal system. That is, demonstrate that the system is consistent, within the finite standpoint.

This is the Hilbertian proposal. Hilbert recognized the essential difficulty of the problem, and tried to deal with the problem from an elevated perspective. This program made it possible to approach the foundations

---

[2] A formalized proposition in this sense is called a formula, and a formalized proof is called a proof figure.

of mathematics as a well-defined mathematical discipline. This established the area as a recognized mathematical subject.

The execution of Hilbert's Program did not progress rapidly, but it did certainly advance. In 1928, "Grundzüge der theoretische Logik" by Hilbert and Ackermann[3] was published. This book gives Hilbert's exposition of the predicate calculus. The completeness of the predicate calculus (that is, logic is completely formalized in the predicate calculus) is proposed as an open problem there. The completeness of the predicate calculus is a problem that should be first taken up in the study of logic. The fact that this was proposed as an open problem for the first time in the book by Hilbert and Ackemann is a clear indication that mathematical logic had not been studied systematically in depth before the Hilbert School.

In 1928, the achievements of the Hilbert School were being prepared to be published in a weighty book "Grundlagen der Mathematik" by Hilbert and Bernays[4]. The program set forth by the book was probably intended to be the following.

First of all, Hilbert's Program, the formalism and the finite standpoint more precisely, was to be stated. Next the predicate calculus, and then an exposition of the ε-calculus, which was a distinctive feature of the Hilbert School, was to be included.

Let me describe the ε-calculus.

Compared with logical connectives of the propositional calculus, $\neg$ (not), $\wedge$ (and) and $\vee$ (or), the quantifiers $\forall$ (for all) and $\exists$ (exists) can be regarded as the notions that are troublesome. Namely, the truth values of $\neg A, A \wedge B$ and $A \vee B$ can immediately be reduced to those of A and B, whereas the truth value of $\forall x A(x)$ cannot be decided unless the truth values of infinitely many propositions $A(0), A(1), A(2), A(3), \cdots$ are decided even for the case where x ranges over the domain of natural numbers, since then $\forall x A(x)$ means

$$A(0) \wedge A(1) \wedge A(2) \wedge A(3) \wedge \cdots.$$

The situation is the same with $\exists x A(x)$. In order to decide whether there is an x satisfying $A(x)$, one has to examine infinitely many propositions

$$A(0), A(1), A(2), A(3), \cdots.$$

---

[3] Springer, Berlin,1928

[4] vol.1, Springer, Berlin, 1934; vol. 2, 1939

The situation being as it is, even when x ranges over the set of natural numbers, when x ranges over the set of all real numbers or furthermore when x ranges over a set which is of much higher cardinality, $\forall x$ and $\exists x$ are highly transcendental notions, and so we might say that they introduce foundationally difficult problems.

Hilbert developed the ε-calculus in order to avoid this problem. Now define $\varepsilon x A(x)$ as

'an element x which satisfies $A(x)$ if there exists such an x'

and call it an ε-term. Then $\exists x A(x)$ is equivalent to $A(\varepsilon x A(x))$, and $\forall x A(x)$ is equivalent to $\neg \exists x \neg A(x)$, and hence we can eliminate $\forall$ and $\exists$ by introducing $\varepsilon x A(x)$. Furthermore, the inferences regarding to $\forall$ and $\exists$ are all derivable from the following single law on the ε-symbol.

ε-**axiom** $A(t) \rightarrow A(\varepsilon x A(x))$

Here t is an arbitrary term.

Of course, to an extent this is, though interesting, essentially a mere rewriting. However, Hilbert devised a method to pursue the consistency of number theory from the standpoint of the ε-calculus.

Suppose we try to prove that no contradiction is derived from

$$A_1(t_1) \rightarrow A_1(\varepsilon x A_1(x)), \cdots, A_m(t_1) \rightarrow A_m(\varepsilon x A_m(x)).$$

We may assume that no free variables occur here. Furthermore, let us consider a simple case where the ε-symbol does not occur in $A_1(a_1), \cdots, A_m(a_m)$. In such a case, we replace an ε-term such as $\varepsilon x A_i(x)$ with a natural number n_i. $t_1, \cdots, t_m$ consist of combinations of ε-symbols and some simple functions like $+$ and $\cdot$, and hence we can compute the values of $t_1, \cdots, t_m$. Let us denote these values by $l_1, \cdots, l_m$. Then the axiom group as above will become of the form

$$A_1(l_1) \rightarrow A_1(n_1), \cdots, A_m(l_m) \rightarrow A_m(n_m).$$

If these axioms are correct, then it is obvious that no contradiction is derived from them, and hence it is obvious that no contradiction appears in the original proof. One of the formulas above, say the ith one, is not correct only if $A_i(l_i)$ is correct and $A_i(n_i)$ is not correct. In such a case, since A_i is constructed by $\neg, \wedge, \vee$ as well as simple symbols such as $\leq, +, \cdots$, one can compute the truth values of all of $A_i(0), A_i(1), \cdots, A_i(l_i)$. Among them, take the first k satisfying $A_i(k)$, denote this k by $\tilde{n}_i$ and let $\varepsilon x A_i(x)$ be $\tilde{n}_i$. Then everything goes well and the consistency proof is completed.

Of course this argument does not go through so easily when the ε-symbol occurs in $A_i(a)$. Even in such a case, one can repeatedly rewrite an ε-term by replacing a term of the form $\varepsilon x B(x)$ in it with some natural number. Then, after some finite steps, one can make a natural number correspond to each ε-term so that the finite number of axioms used in the proof are all satisfied. The only problem that remains is how to show it from the finite standpoint. One has only to discover an appropriate process of replacement with natural numbers or of rewriting. This was Hilbert's way of thinking. This approach exhibits vividly the fundamental ideas intended by Hilbert on the finite standpoint and consistency proofs from that standpoint.

I have not written how to treat mathematical induction; mathematical induction is equivalent to the well ordering principle. In the rewriting of ε-terms above, an $\varepsilon x A(x)$ can be in fact treated as 'the least number x satisfying $A(x)$' and so it should be clear that success of the above method would entail the consistency of number theory with mathematical induction.

For this reason, the Hilbert School applied its efforts to the study of the ε-calculus. Hilbert and Bernays first studied some fundamental properties of the ε-calculus, and then Ackermann published a consistency proof of number theory with this method, although there was an error in his proof. The first partial success was by von Neumann in 1927, and he gave a consistency proof, using the same method, of a weakened system of number theory in which the mathematical induction is restricted to the propositions without $\forall$ and $\exists$.

6.3 Über das Unendliche

It is interesting to see how the Hilbert School thought about the following problem.

In 1925 Hilbert published an article entitled "Über das Unendliche" in Mathematische Annalen. In this article, Hilbert first considered the totality of all well-ordered sets of natural numbers, then considered transfinite ordinal numbers which can be expressed in terms of these well-ordered sets, and last showed that the totality of all subsets of natural numbers can be defined constructively by using these transfinite ordinal numbers. He developed a program to solve the continuum hypothesis affirmatively using this result.

It can be safely said that this program was an important focus of the Hilbert School at the time. Indeed, Hilbert stated in the introduction to the previously mentioned book of 1928 by Hilbert and Ackermann that it was intended to set the stage for the book by Hilbert and Bernays to be published later, and described Hilbert's Program as above and "Über das Unendliche" as preliminary to the book by Hilbert and Bernays.

In this plan Hilbert referred to:

'the totality of all well-ordered sets of natural numbers which can be defined primitively recursively'

which should have been:

'the totality of all well-ordered sets of natural numbers which can be defined constructively.'

In 1928, Ackermann proved that with this definition of the totality of well-ordered sets, one could not succeed. Nevertheless, this was a natural mistake when it is understood that the area was still in an early stage of development. Hilbert's plan was interesting all the same. His plan was such that, if at an early stage some aspect would not work, it could be modified and studied further so that fruitful results could be obtained. How far had Hilbert's plan progressed by 1930? Although it is an interesting question, there is no way of knowing now. (I wish I had asked Bernays about it while he was alive. I regret that there is no longer anyone left who knows.)

6.4 The Hilbert School in 1930

In 1930, young Herbrand, then 22 years old, joined the group surrounding Hilbert, and published an article on a deep theorem, known as Herbrand's resolution theorem. This theorem more or less corresponds to Gentzen's fundamental theorem published later in 1934, bringing research on the predicate calculus to present day standards.

This does not really mean that Hilbert's Program had made a great strides. Nevertheless, with his group of young students, Bernays, Ackermann, von Neumann, Herbrand, and in addition the 20 year old Gentzen, Hilbert's cadre must have been confident of their success, anticipating a bright future.

6.5   Gödel enters the arena

In 1930, Gödel then 24 years old proved the Completeness Theorem of the predicate calculus in

"Die Vollständigkeit der Axiome des logischen Funktionenkalküls," Mh. Math. Phys.

Of course, this theorem affirmatively resolved a conjecture which Hilbert and Ackermann had presented as an open problem, and so we can say that it conforms to the Hilbert line. However, immediately thereafter, Gödel proved his famous Incompleteness Theorem in "Über formal unentscheidbare Sätze der Principia Mathematica und verwandter Systeme," Mh. Math. Phys. in 1931. This was a sensational result. Its meaning resides in the following two theorems.

(1) For a consistent axiom system which contains number theory, there is necessarily a proposition which is undecidable in that system. Namely, there is a proposition such that neither it nor its negation is provable.

(2) A proposition which is a formalization of the consistency of a consistent axiom system which contains number theory cannot be proved within this system.

The first theorem was earthshaking. At that time, not only Hilbert but everyone else believed beyond doubt that, for any proposition of number theory, either it or its negation could be provable in, for example, set theory. Because it was assumed that this would be the case, it was not even explicitly stated. The first theorem of Gödel claims that the extreme opposite was the truth. The Hilbert School, the Vienna School, $\cdots$; the whole world was stunned.

The second theorem was a severe blow to the Hilbert's Program. The inferences of the finite standpoint are simple and elementary. There are hardly any inferences which cannot be formalized in number theory. This entails that it is almost impossible to execute a consistency proof of number theory within the finite standpoint in any practical sense. Gödel himself writes at the end of this paper, however, that his results do not completely exclude the possibility of carrying out a consistency proof from Hilbert's finite standpoint. Also, Hilbert himself writes in the introduction of Hilbert-Bernays (Grundlagen der Mathematik I) published in 1934 to the effect: 'The opinion that "my program on the foundations of mathematics was ruined by Gödel's theorem" has turned out to be completely erroneous. Gödel's theorem asserts merely that in consistency proofs one has to use the finite standpoint in a sharper way.' We might say that this wish of Hilbert's was realized by Gentzen's consistency proof of number theory in 1936. However, as Hitler came to power in 1933 and forced Jewish people out of the universities, the mathematics department at Göttingen disintegrated. At the same time, due to the Incompleteness paper by Gödel, Hilbert's Program was devastated.

The next big event in mathematical logic also happened in the 1930's and involved Gödel and Hilbert.

In 1938, Gödel proved in his

"The consistency of the axiom of choice and the generalized continuum hypothesis," Proc. Nat. Acad. U.S.A.

that the axiom of choice and the generalized continuum hypothesis are consistent with set theory. We could say that this article was the starting point for modern set theory. Incidentally, Gödel made use of Hilbert's idea in "Über das Unendliche." Exaggerating slightly, we can say that the difference between the ideas of Gödel and of Hilbert is that, while Hilbert considered constructible well-ordered sets, Gödel considered all well-ordered sets. Obviously Gödel got the main idea of his article from Hilbert (as well as from the ramified hierarchy of Russell and Whitehead).

6.6   The disintegration of Hilbert School

A good case can be made that Gödel's three great achievements in the 1930's were the major results in mathematical logic of the period. After this, the Second World War broke out, and Hilbert died in 1943.

There are several reasons why the Hilbert School disintegrated. The first was Gödel's work. The second was the persecution of the Jewish people by Hitler. The third was the premature death of Herbrand. What would have happened without Gödel's work, Hitler's appearance and Herbrand's premature death?

There is no doubt that then the Hilbert School would have prospered and progressed. One way or another, Herbrand and Gentzen would have completed the consistency proof of the theory of natural numbers. As it was Gentzen did achieve the consistency proof of the theory of natural numbers in 1936 as I have mentioned earlier, with an apparent indirect influence of

Gödel's Incompleteness Theorem in his treatment.

On the other hand, let us consider the direction of Hilbert's "Über das Unendliche." Shortly after Hilbert applied primitive recursive functions to the problem the notion of recursive functions was conceived. One of the approaches was due to Herbrand and Gödel. Had recursive functions been available, Hilbert's plan could have been carried forward. It would not necessarily have reached Gödel's result on the continuum hypothesis, but it is certain that interesting results could have been achieved. My view is that such a plan did not bear fruit in the Hilbert School because most of the Hilbert School ceased to exist due to Gödel's theorem and those who remained were disheartened.

On the other hand, what were Gödel's later accomplishments? Although Gödel published several interesting works after those of the 1930's, none of them are comparable with the three mentioned above.

I have written about the 1930's in this section mainly from the standpoint of the Hilbert School. Why is it that Gödel's major works were concentrated during this period and, furthermore, that they are deeply related to those of Hilbert? In "Hilbert and Gödel" (Chapter 3), I presented a guess by spotlighting Gödel. What will be the reader's view?

Chapter 7

Axioms of Arithmetic and Consistency —
The Second Problem of Hilbert

In order to state the Second Problem of Hilbert [1], I will first explain about contradictions that were discovered in set theory.

In the latter half of the previous century[2], G. Cantor proposed the notion of sets, and created set theory, or naive set theory in contemporary terminology.

Cantor's notion of sets facilitated abstraction and, based on that, abstract arguments of high quality became possible.

In that sense, Cantor's set theory was a necessary precondition to the conceptualization and development of modern mathematics. On the other hand, Cantor's set theory derives its power from the fact that one can judge a proposition intuitively. Notwithstanding its intuitivity, a contradiction was derived from Cantor's set theory.

The first contradiction (referred to as a paradox during the formative period of set theory)[3] was discovered by Cantor himself in 1895, although it was not made public. In the next year 1896, however, Cantor communicated it to D. Hilbert.

Hilbert was a great mathematician of deep insight. He always gave consideration to the essence of a problem, and it was his characteristic first to see through to the core of a problem and then to resolve the problem from a more general point of view. To Hilbert, the concept of set, in terms of which one could deal with abstract notions freely, was irreplaceably valuable in mathematics. I believe it can be claimed that the spirit of the times which created the mathematics of the first half of the 20th century arose from the notion of sets as well as from Hilbertian formalization.

---

[1] The compatibility of the arithmetical axioms:1900

[2] 19th century

[3] Known also as Burali-Forti paradox, caused by admitting the collection of all ordinals as a set.

Yet, from this very set theory, a contradiction arose. Hilbert was afraid more than anything else that modern mathematics, which he believed in, might be damaged due to this contradiction. It is said that Hilbert's state of mind at that point was: "How could I abandon so easily the paradise that Cantor created!"

What could he do? Hilbert proposed the Second Problem! We might understand that this was his way of raising a question in such a difficult situation.

Previously in "Grundlagen der Geometrie"[4] Hilbert proved the consistency of geometry by defining a model of geometry in terms of real numbers. This work reduced the consistency of geometry to the consistency of the theory of real numbers, which does not fundamentally solve the problem.

The fact that he raised the issue of the consistency of arithmetic, the most fundamental theory in mathematics, indicates that Hilbert thought that the problem of the consistency of set theory was a fundamental and serious problem. Hence one ought not expect an easy and superficial solution. Thus one should start with the problem of arithmetic, the most basic part of mathematics.

For this Second Problem, Hilbert proposed the following plan, known as Hilbert's Program, and began to devote himself to its investigation.

(1) Formalize mathematics and treat it as a formal system.
   For this reason, Hilbert's standpoint is called formalism.
(2) In this formalized system, a proposition is expressed with a string which is composed of some special symbols which stand for basic propositions, variables and logical notions respectively. Mathematical inferences and proofs are all described as concrete rules which apply to these strings, which are clusters of symbols as above. [5]
(3) Only the arguments which are applied to the formalized proofs that are concretely considered as above and which are simple and reliable can be admitted. Namely, the strings which are relevant here are concrete strings consisting of finitely many symbols. This standpoint is called the finite standpoint.
(4) Complete the foundational underpinnings of the formalized mathematical system by proving that no contradiction can be derived in this formal system. That is, demonstrate that the system is consistent,

---

[4] 1899

[5] A formalized proposition in this sense is called a formula, and a formalized proof is called a proof figure.

within the finite standpoint.

Under the influence of Hilbert, gifted people such as Bernays, Ackermann, von Neumann, Herbrand and Gentzen pursued Hilbert's Program. By 1930, Hilbert and the Hilbert School had achieved the following.

- The study of logic, that is, of the predicate calculus, had been completed. Namely, it had reached our present stage of knowledge.
- The consistency of an extremely weak system of number theory had been proved by von Neumann, namely the system of number theory in which the application of mathematical induction is restricted to formulas without ∀ and ∃.

Then in 1931 came what we know as Gödel's Incompleteness Theorem. The Incompleteness Theorem actually consists of the following two theorems.

- For a consistent axiom system which contains number theory, there is necessarily a proposition which is undecidable in that system. Namely, there is a proposition such that neither it nor its negation is provable.
- A proposition which is a formalization (in the language of arithmetic) of the consistency of an axiom system which contains number theory and which is consistent cannot be proved within this system.

Although the first theorem is not particularly related to the Second Problem of Hilbert, it is a profound result with influence on the whole of mathematical logic. It was a great and remarkable theorem which had not previously been imagined.

It was the second theorem that was fatal to Hilbert's Program. An inference which arises from the finite standpoint is basic and elementary. An inference which cannot be formalized in arithmetic cannot be conceived of from such a standpoint. In that sense, we can say that Gödel's second Incompleteness Theorem asserts that Hilbert's Program is almost unrealizable. However, Gödel writes in the last part of his article that his results do not completely deny the possibility of Hilbert's Program. Hilbert too writes in the introduction of "Grundlagen der Mathematik I" by Hilbert-Bernays as follows.[6]

---

[6]translated from German by the translators

In consideration of this goal I would like to suggest that the currently fashionable opinion, that the consequence of Gödel's recent results is that the program of my Proof Theory cannot be carried out, is shown to be erroneous. These results indeed only show that for the extension of the consistency proof it is necessary to apply the finite viewpoint in a sharper manner than would be necessary for the consideration of a simpler formal system.

Indeed, G. Gentzen in 1936 accomplished a consistency proof of number theory. But here the interpretation of the finite standpoint was certainly extended. That is what Hilbert meant by 'in a sharper manner.'

Gentzen's consistency proof of arithmetic was an important result for Hilbert's Program or at least his finite standpoint. Let me therefore discuss Hilbert's phrase 'in a sharper manner' a bit more. In Gentzen's proof, an equivalent of transfinite induction up to ε_0 is applied. Otherwise every step is finite and elementary, which causes no problem. Thus, I will focus on transfinite induction.

First, natural numbers are constructed from the finite standpoint. For example, we define numbers as the expressions of the form $1, 1 + 1, 1 + 1 + 1, \cdots$. Precisely speaking, this is done by recursive definition. Further, with regard to these natural numbers, order relation, addition, multiplication and their basic properties can be defined and proved from the finite standpoint in a strict sense.

Then for a proposition $A(n)$ on the natural number n, which has been defined from the finite standpoint, the principle of mathematical induction holds. Namely, if it is known that $A(1)$ holds and that $A(n)$ implies $A(n+1)$ for all n, then $A(n)$ holds for every n.

This does not mean that one admits mathematical induction as a basic principle, but it is a self-evident proposition since one eventually reaches $A(n)$ for any n if one proves $A(m + 1)$ from $A(m)$ successively. That is, it becomes a self-evident principle by the basic approach of the finite standpoint to arguing concretely on concrete strings of symbols presented in front of one's eyes.

Next let us consider the main subject, transfinite induction up to ε_0. Here ε_0 is the ordinal number of the form

$$\varepsilon_0 = \omega^{\omega^{\omega^{\cdots}}}$$

In order to deal with ordinals up to ε_0 from the finite standpoint, we must first assume that natural numbers are $0, 1, 2, 3, \cdots$, not $1, 2, 3, \cdots$ as before. For this purpose, of course we only have to add one symbol 0.

With this preparation, in order to construct the system of ordinal numbers below ε_0, it is sufficient to introduce a new letter ω, and define operations $\alpha + \beta$ for ordinals α and β (As an operation on strings of symbols, this means to construct, from $\alpha = \mu_1 + \cdots + \mu_n$ and $\beta = \nu_1 + \cdots + \nu_m$, the new string $\mu_1 + \cdots \mu_n + \nu_1 + \cdots \nu_m$.) and ω^α (This means literally to write ω^α [7].). Then, the order relation on the system of ordinal numbers below ε_0 and basic properties concerning two operations above can be developed rigorously according to the finite standpoint in the narrow sense.

We introduce the notion that "α is accessible." The rigorous definition of this is that every (strictly) decreasing sequence of ordinal numbers starting from α is finite. [8]

Now, given a natural number n, every decreasing sequence starting from n terminates within finite steps. We can thus recognize that a natural number n is accessible. From this, we can understand that the notion that a natural number n is accessible is a clear notion. There is not much reason to oppose this idea by claiming that the notion that all decreasing sequences terminate within finite steps is a Π_1^1 notion in Kleene's hierarchy. What is important is not which hierarchy the notion belongs to, but how clear it is.

From the discussion above, it can immediately be understood that ω is accessible. By continuing the same argument, we will know that $\omega \cdot 2 (= \omega + \omega)$, $\omega \cdot 3 (= \omega + \omega + \omega)$, $\cdots$ are accessible. From these we will know that ω^2, similarly $\omega^3, \omega^4, \cdots$ and further ω^ω are accessible. If we continue this reasoning, we will know that $\omega_2, \omega_3, \cdots$ (defined by: $\omega_1 = \omega, \omega_{n+1} = \omega^{\omega_n}$) are accessible . Therefore, we will know that ε_0 is accessible.

If I may add one word, Gentzen has shown that, while accessibility of ω_n for each $n = 1, 2, 3, \cdots$ can be proved in the system of Peano arithmetic PA[9], accessibility of ε_0 cannot be proved in PA. However, accessibility of ε_0 is obviously equivalent to the proposition "ω_n is accessible for all n from the finite viewpoint." This fact is an exemplary case in which the finite standpoint can produce a stronger result than a formalized system by restricting the objects it deals with to concrete strings of symbols, concrete operations and notions which can be examined clearly from the finite standpoint.

As for myself, I believe that, once a good system of ordinal numbers

---

[7] Literally write α on the right shoulder of ω without assigning particular meaning.

[8] More precisely, there must be a concrete method to show that such a sequence terminates.

[9] the formalized system of number theory

has been constructed within the finite standpoint, its accessibility proof can be carried out within the finite standpoint once the properties of that system will have been sufficiently studied from the finite standpoint. It will be certain that the more powerful a system of ordinal numbers be, more objections to it based upon the Π_1^1 concerns and the like as mentioned above there will be. I believe, however, that the fundamental idea of the finite standpoint transcends these objections.

I will list below some major events concerning Hilbert's Program after Gentzen.

Although Hilbert proposed to study consistency problems of arithmetical axioms, there is no doubt that his real objective was the consistency of the theory of real numbers, of analysis, and further of set theory.

The theory of real numbers can be reduced to the theory of sets of natural numbers. Analysis can be reduced to the theory of finite accumulations of notions of sets over natural numbers, that is, the set of subsets of natural numbers, the set of subsets of the previously defined set, and so on and so forth. Therefore, the consistency problems of the theory of real numbers and of analysis were the starting points for the study of the consistency problem of set theory.

Since the notion of sets can be regarded as a mathematical interpretation of propositions, analysis can be regarded in its essence as the theory of natural numbers augmented by the predicate calculus of higher order, and the theory of real numbers can be regarded in its essence as the theory of natural numbers augmented by the second order predicate calculus.[10]

I extended Gentzen's logical system LK[11] to define a system GLC, and proposed my fundamental conjecture that the cut elimination theorem [12] holds also in GLC. I then proved that if my conjecture indeed holds in GLC, then the consistency of analysis follows, and that if it holds in the second order subsystem of GLC, called G^1LC, then the consistency of the theory of real numbers follows.

I conjectured that the resolution of my fundamental conjecture from the finite standpoint would be the central problem of this field. I planned, assuming its success, to propose mathematical open problems, including consistency problems, and to reconsider the consistency problem in broader context, in order to break the stagnation of research on impredicative pred-

---

[10]the predicate calculus with quantifiers over sets

[11]a formulation of predicate calculus

[12]This is a theorem which claims that any proposition provable in a logical system can be in fact proved without detour. It holds in LK.

icate logic [13] due to Gödel's Incompleteness Theorem. Of course it was an extremely difficult departure to study the fundamental conjecture for a higher order subsystem of GLC from the finite standpoint. Even with the second order system G^1LC I was unable to make a breakthrough, although I made serious efforts. What I achieved was a solution of the fundamental conjecture on the Π_1^1-comprehension axiom and as its extension a consistency proof of a strong inductive definition. For this purpose, I developed the theory of ordinal diagrams, a system of some ordinal numbers which is much stronger than Gentzen's ε_0, from the finite standpoint. In fact this theory of ordinal diagrams is natural since it expresses the process of the cut eliminations. It can thus be justified from the finite standpoint. To the question in what kind of the finite standpoint we can carry out the accessibility proof, only a rough outline of its answer has been published so far. My excuse is the following. All the results I have obtained so far can be situated only halfway up the road to the resolution of the fundamental conjecture on GLC, and I have not been in the mood for examining half done proofs in detail. Having resolved the fundamental conjecture on the Π_1^1-comprehension axiom, I continued to establish the cut elimination theorem of the provably Δ_2^1-comprehension axiom in collaboration with Mariko Yasugi. We further determined the ordinal number associated with the Δ_2^1-comprehension axiom, applying a result by Friedman and a result concerning the inductive definition. H. Friedman's result uses a method of nonstandard arithmetic, hence cannot be carried out in Hilbert's finite standpoint. Since our intention originates in Hilbert's Program, I did not feel that our last result had resolved the case of the Δ_2^1-comprehension axiom.

My fundamental conjecture has been proved to hold independently by Motoo Takahashi and Dag Prawitz. Their proof methods are, however, set theoretic, and so their works cannot be regarded as execution of Hilbert's Program.

After this period, proof theory has been developed by the Schütte School. One characteristic of the Schütte School is that they shelve the troublesome problem of Hilbert's Program, which is entangled with Gödel's theorem, and confine themselves to the theme of determining the proof-theoretic ordinals of formal systems. In this sense, the above mentioned result of ours has resolved the case of the Δ_2^1-comprehension axiom.

Such being the circumstances, I, who entered this field through Hilbert's

---

[13] second order predicate calculus without an restrictions

Program, view the works of the Schütte School as follows.

(1) Since Hilbert's Program is a difficult problem, it may be necessary to consider the problem by reducing it to their approach.
(2) If the work of the Schütte School is completed, then some day it will be useful to Hilbert's Program.
(3) What sort of significance does it have if it cannot be linked to Hilbert's Program?

It seems to me that one needs a clarification of the statements above.

The works of the Schütte School had been promoted by Schütte, Pohlers and Buchholz. Recently Michael Rathjen and Toshiyasu Arai have independently resolved the case of the Π_2^1-comprehension axiom. Notwithstanding that I am critical of this direction as stated above, their work is a blessing and a wonderful achievement. Having heard Rathjen's talk only once, I cannot comment in detail. What I can explain is that their proof-theoretic ordinal is a miniature of a large cardinal whose existence can be assured only by a new axiom of set theory. On the one side, one could say that, due to their method, the relation between set theory and the foundations of mathematics has become closer, though I feel that it has moved away from Hilbert's Program. Is this what Gödel's Incompleteness Theorem means?

I am currently devoted to the problem of determining bounds of weak systems of arithmetic, and believe that this is a serious issue for the foundations of mathematics. This can be called the dual of Hilbert's Program, and in this area even Gödel's Incompleteness Theorem is powerless.

Chapter 8

A Report from Gödel '96

Gödel was born in 1906, and so this year (1996) a conference entitled "Gödel '96" commemorating the 90th anniversary of Gödel's birth was held in Brno in the Czech Republic.

I attended this conference with great anticipation. Among many logicians, Gödel is the only one whom I regard as a genius. Discussions and conversations I enjoyed with him over the years were precious experiences for me. Incidentally, although I am counted as one of those who were personally acquainted with Gödel, I know only a tiny bit of one side of Gödel, and I cannot possibly claim a complete understanding of Gödel's world view from our discussions.

The editing work of "*Kurt Gödel, Collected Works*" [1] steadily progressed; the first volume was published in 1986 and the second volume was published in 1990. Furthermore, many notes and other documents have been investigated in some detail.

I even expected that I might hear interesting new facts at the conference which relate to a variety of subjects of discussions between Gödel and myself.

Perhaps because of too great expectations, this conference was unsatisfactory to me. I will take up only a selected few of the talks at the conference, and present my personal view of them.

8.1 A program on a new axiom

The opening talk was Solomon Feferman's "Gödel's program For New Axioms: Why, Where, How And What?" My hopes for this talk were enor-

---

[1] Volumes I~III, OUP

mous. Gödel's thoughts on new axioms excited the community intensely. Gödel was trying to resolve the continuum problem in terms of large cardinals on the one side, while he attempted to resolve it in terms of an axiom called the square axiom[2] on the other. This was a subject that came up time and again in my conversations with him. I gather that Solovay is responsible for this part of Gödel's notes. As I was once asked questions by Solovay on some parts of Gödel's manuscripts, I was curious about Gödel's thoughts, and so I had hoped that they would be elucidated in this talk according to recent investigations.

Now what Feferman called "Gödel's new axiom" had nothing to do with Gödel. Feferman talked on his work on a system which is obtained from a system of arithmetic by repeated addition of the reflection principle, and it is reasonable to claim that it had nothing to do with Gödel. Adding a reflection principle is a slight extension of adding a consistency statement. Since consistency cannot be derived in the original system, the system thus extended is stronger than the original system. This fact itself is a consequence of Gödel's Incompleteness Theorem. Considering this, it is not unrelated to Gödel, but it belongs to a different category from what Gödel called a new axiom. Sadly, I could not find anything I had expected.

The next talk was "Forcing on Bounded Arithmetic" by G. Takeuti and M. Yasumoto. I would like to take this opportunity to explain our article a little bit. It was Paris and Wilkie who first applied the forcing method (a method which Cohen had developed in order to prove independence and consistency of some propositions in set theory) to weak systems of arithmetic, or bounded arithmetic.[3] Although results of the work by Paris and Wilkie are somewhat as expected, Ajtai followed them in publishing excellent works on the pigeon hole argument and enumeration modulo p. This stimulated general interest and quite a few articles have been published on improvements and variations of Ajtai's work to this day.

We presented a kind of forcing entirely different from the forcing mentioned above. First, our forcing applies the method of Boolean valued models and uses large Boolean algebras. Ajtai's forcing is in fact an idea resembling forcing and cannot be called forcing in a strict sense. In that sense, our forcing can be said to be parallel to the forcing in set theory. This fact implies the following difference. With his forcing method, Ajtai

---

[2]See Appendix A.
[3]first proposed by S.Buss

proved that, in the hierarchy of computational complexity

$$AC^0 \subseteq NC^1 \subseteq AC^1 \subseteq NC^2 \subseteq \cdots \subseteq N \subseteq P \subseteq NP \subseteq \cdots,$$

various low hierarchies between AC^0 and $AC^0 + \varepsilon$ are mutually distinct. Here $AC^0 + \varepsilon$ is situated very near AC^0 and far below NC^1. I feel that this method does not reach NC^1 no matter how one improves it. Namely, the existing forcing is a theory closely tied to AC^0. In our case, Boolean algebras are colossal and our theory is developed parallel to set theory. For this reason, our forcing works efficiently, and it has a close relationship to the inequality $P \neq NP$.

It has been some time since the problem of $P \neq NP$ was thrown into the limelight. Although during this time many brilliant people have attacked the problem, nobody has succeeded in mastering it. The idea that the existing approach of applying the diagonal argument to the combinatorial method is futile and that a theory of higher-degree must be necessary is spreading. There are only two examples of successful separation of two systems in mathematical logic, that is, the Incompleteness Theorem and forcing (implying respectively separation between systems of number theory and of analysis and separation between set theory and set theory augmented by the continuum hypothesis). In particular, forcing resolved beautifully a problem which the diagonal argument, represented with Gödel's Incompleteness Theorem, was unable to solve. I suspect that $P \neq NP$ is also such a case. Our theory was developed in order to attack $P \neq NP$. I seriously believe that, if somebody understands this theory and comes on an appropriate idea, then $P \neq NP$ will be resolved all at once.

Let me note a connection of our theory to Gödel. In a letter to von Neumann in 1956 Gödel mentioned a problem which was closely related to the $P \neq NP$ problem. I cannot help thinking that, had the study of computational complexity started in 1956, Gödel might have accomplished great things in this field.

Gödel also felt attached to the method of Boolean valued models. The reason is that he was trying to develop a method of systematically re-interpreting logical symbols in order to resolve independence problems. For example, he was attempting to prove the independence of the axiom of choice from Zermelo set theory. However, as one can understand from his article in Dialectica, in which he proved the consistency of number theory, re-interpretation of logical symbols tends to produce confusion and to get out of hand, and hence Gödel was discouraged. The method of Boolean models was a model theoretic, simple and direct variant of his theory of re-interpretation. In passing, when I developed Boolean valued analysis by the method of Boolean valued set theory in terms of the Boolean algebra of projection operators over Hilbert spaces, Gödel invited me to the Institute in Princeton for a week, and he encouraged me by discussing the subject extensively. I myself believe that, if Gödel were alive now, he would be interested in our present work.

8.2 Gödel's program on the continuum

The next invited talk I would like to take up is the one by Magidor. This conference was a strange one. It was scheduled August the 26th through the 29th, and the deadline for manuscripts of talks was April the 28th (Gödel's birthday), and Proceedings of talks entitled

> Petr Hájek (Ed.), Gödel'96, Logical Foundations of Mathematics, Computer Science and Physics — Kurt Gödel's Legacy, 1996, Lecture Notes in Logic 6

was distributed at the conference. In connection with this, there was an incident concerning Parsons. Parsons had been invited in April to a conference at Keio University [4], but he was unable to come. The reason was that the deadline for the manuscript of Gödel'96 was nearing and, as he was struggling with his manuscript in order to meet the deadline, he was unable to visit Japan.

Now the title of Magidor's talk was "title to be announced" and so we did not know its content until the beginning of his talk. His article is certainly not included in the Proceedings of Gödel'96. I cannot write down the proper title as I only heard the speaker say it once at the beginning, but it was more or less, Gödel's program on the Continuum Problem was a Total Failure.

The talk itself was not quite as negative about Gödel's program as one might guess from the title. His assertion consisted of the following.

(1) Although Gödel expected that the continuum problem would be resolved by axioms of large cardinals, the axioms of large cardinals known to the present do not resolve the continuum problem.

(2) Gödel hoped that, aside from the axioms of large cardinals, an axiom would be discovered with qualities such that its consequences would

---

[4] In Tokyo

make us accept its validity (The validity may be explained in terms of its relation to physics.) and that it might resolve the continuum problem. At present there is nothing like that. (Here he listed axioms such as Martin's axiom which failed to do the job.)

Incidentally, he came to me the day before his talk to ask me about my work relating to Gödel's square axiom. Since the schedule of the conference was tight, we could not find time for discussion. Fortunately I had a reprint of my paper, which I could hand over to him.

In his talk, he mentioned recent work of Todorčević which had something to do with mine. It seemed to me that Todorochevich's idea was related to Gödel's idea as well as to the Hausdorff axiom of continuity (This is not an axiom proposed by Hausdorff himself, but is an axiom to which Gödel assigned the name), which Gödel always referred to, although Todorčević came up with the idea independently of Gödel. I asked Magidor later about it, and he promised to send me a copy of his report on the subject. I am seriously interested, and I look forward to receiving a copy; I have not received it yet.

8.3   On Hao Wang

The next invited talk I shall take up is
   "Hao Wang as Philosopher and Interpreter of Gödel"
by Charles Parsons, whom I mentioned earlier. Hao Wang was extensively discussed in this conference. One reason for this was of course that Hao Wang was close to Gödel and was well acquainted with Gödel's philosophy, but a decisive reason was that Hao Wang was the first president of the Kurt Gödel Society, which was one of the sponsors of the Conference, and was recently deceased. So this conference was also meant as his memorial service. The second president of the Kurt Gödel Society was R. Jensen, and the third president is D. Mundici.

Parson's talk centered on four works by Hao Wang:

"*From Mathematics to Philosophy,*" London: Routledge and Kegan Paul, 1974.
"*Beyond Analytic Philosophy, Doing Justice to What We Know,*" Cambridge, Mass: MIT Press, 1985.
"*Reflections on Kurt Gödel,*" Cambridge, Mass: MIT Press, 1987.
"*A Logical Journey: From Gödel to Philosophy,*" Cambridge, Mass:

MIT Press. (forthcoming)[5]

To my regret, I, who do not understand philosophy, could not feel excited with this talk.

8.4 Gödel's theory of relativity

The last invited talk I am going to take up is

George Ellis: Contributions of K. Gödel to Relativity and Cosmology.

This talk was interesting. Unfortunately I cannot explain it in detail, as I am not acquainted with the general theory of relativity. What I understood of the major contribution of Gödel included in this talk was the following. Gödel proved that the fundamental ideas of relativity at that time, that is, what had been generally agreed as to the topology of the space which could be derived from the equations of relativity and the belief that the law of causality would follow as a consequence, were all completely erroneous. As a result, much excellent research followed from this on these two important conditions. Among them is important work by Penrose, Hawking and Raychaudhuri. We would be justified in considering that Gödel's work can be located as the starting point for all these researchers. I got the impression that wormholes, black holes and time warps are related to this work.

In passing, among Gödel's articles, there are quite a few in which only the conclusions are stated without proofs (the second paper[6], for example), and it is quite a hard job to execute the proofs, even for a specialist.

It seems that Gödel derived the conclusions above by elaborate computations.

This was a side of Gödel which I had not known. Besides, it was a real surprise to me to learn of Gödel's accomplishments even in the area of relativity, which went much farther than I had imagined. I would strongly recommend this article to the reader.

Let me add one more thing. There was a sentence at the beginning of this talk that Gödel obviously had come to this work through the influence of his friendship with Einstein. Certainly that must be the case. However, I would be interested to find out how strong this influence was.

Once in Princeton I heard a story as follows. Gödel had began to work on the theory of relativity before he met Einstein. Someone heard

---

[5] published since

[6] of relativity: There are three papers by Gödel concerning the theory of relativity.

Gödel talking about his idea, and asked him: "Have you mentioned it to Einstein?" When Gödel replied: "No, because nobody has yet introduced me to Einstein," the questioner introduced Gödel to Einstein. Furthermore, Gödel himself told Kreisel: "When I was a student, I hesitated over which to choose as my major, the theory of general relativity or mathematics." Also, Menger, who had known Gödel when he was young, has reported that Gödel in his twenties studied the relationship between the theory of general relativity and the ideas of Hegel. Taking all this into account, we might guess that a significant part of his first paper[7] had been worked out before he befriended Einstein.

8.5   Gödel in Brno

Gödel was born in Brno (then Brünn). That is the reason why the Conference was held there.

I did not do any sightseeing in Brno. I only went to see the house where Gödel had been born, together with some friends including Baaz, Krajíček and Pudlák. It looked rather dilapidated. My friends jokingly told me that it could be obtained with 10 million yen, and asked me if I would not buy it, move it to Japan and use it for tourism or business. They even went further to a worse joke: "You should live in it yourself. Then we will advertise that this is the house where a great logician was born and where another great logician is now living."

There was a plaque on the wall, on which is engraved that this is the birthplace of Gödel, a mathematician, logician and philosopher.

On the poster and other handouts from the Conference, there was a photo of Gödel in his youth. In fact in the first edition of this book, there was a photo with the title "Gödel at the time of his proof of the Incompleteness Theorem (25 years old)." I simply took the word of a person from Vienna who had given me this photo. On the other hand, there is a caption for the same photo in Gödel's "*Collected Works*," from which I get an impression that it is a part of a souvenir picture of students. Susumu Hayashi pointed out this gap. Looking at the photo with this in mind, Gödel appears much too young. I had expected I could find somebody who would know the truth, but I did not meet anybody who knew much about it. The caption of the photo in the Conference reads: Kurt Gödel in der Zeit seines Aufenthaltes in Brünn. I have almost forgotten German, but

---

[7]of relativity

I understand that Aufenthalt means the time when Gödel visited Brünn having been in Vienna. Then, this photo is indeed a souvenir picture of students. After all, I cannot judge which is true.

Having attended Gödel'96, I have recovered various memories of the past. In particular, I have begun to feel like working on the continuum problem once again and continuing the mathematical conversation with Gödel on the matter in my mind. The Gödel whom I visualize is the Gödel of old, and he is still twenty years older than my present self. It is curious as, in fact, Gödel when I knew him was younger than I am now.

Chapter 9

Having Read "Gödel Remembered"

The book
"Gödel Remembered" Bibliopolis, 1987, Napoli
is a collection of the articles presented at the Gödel Symposium in Salzburg in 1983.

I had been invited to give a talk at this Symposium, but circumstances forced me to miss it. After that I was invited to write an article for the volume, but I had to decline the offer.

In fact, "Gödel Remembered" was the title of my planned talk in Salzburg. I was told that, since I had failed to give my talk there, Weingartner, an editor, took the title and named the book after it.

As I read it, having received a copy of the book, I found it quite interesting. It is a book which I wish those who are concerned with Gödel would become interested in. I would like to introduce the content here, especially for the sake of Japanese, among whom the book has not been widely known.

The first article is an interview with Gödel's older brother Rudolf, entitled "History of the Gödel family." He talks about his family, with Gödel as the central figure. It includes the following facts. Gödel grew up in a wealthy family. Gödel's mother was a good pianist, and she regretted that her children did not show much interest in music. She was of a sunny and lively disposition. When she visited her son in America in her later years, she was pleased with the beautiful yard at his house. (For Gödel too, it must have been a happy period in his life.) Gödel himself did like to listen to music; he was fond of light operas, especially Viennese operetta. To me this article is the most interesting part. There are also many interesting incidents which could be recounted only by a brother. At any rate, I was totally absorbed by it.

The second article is "Remembrances of Kurt Gödel" by Olga Taussky-Todd. Taussky was Gödel's classmate in his Vienna days, and her article describes vividly the atmosphere of the university in Vienna at the time. According to her, Gödel was fond of girls. When Gödel was to depart for Princeton, his friends went to the station to see him off. Gödel boarded the Orient Express and the friends waved farewell. However, Gödel became feverish before he could board the steamship, and so he returned home (without informing his friends) and later started again. There are also stories about the then young Wirtinger, Furtwängler and others, which give us a sense of those long ago days. It has also drawn my attention that she writes about Professor Shokichi Iyanaga and a statistician Midutani (Mizutani?) among the visitors at a tea party at the Vienna University. In one of the photos in the book, we see a couple, who look like Japanese, together with Karl Menger, Taussky and Gödel. I wonder who they are. (Are they the Mizutanis?) I would appreciate it if somebody could tell me who they are. At the end, some conflict between Gödel and Zermelo is mentioned.

The next article is Gödel's impression on students of logic in the 1930s by Stephen C. Kleene. His writing is very typical of Kleene. Those who have an interest in the history of mathematical logic should not miss this article.

The last article is "Gödel's excursions into intuitionistic logic" by Georg Kreisel. This is a big article occupying 122 pages of a booklet of 186 pages. It has played a substantial role in giving the booklet the style of a book. Typically for Kreisel, it contains too much information in various forms. It is a bit dense to read through, even for me, whose speciality is close to his. Therefore, I read here and there, looking for parts of interest to me. One is Gödel's view of Gentzen, in which can be read:

He often called Gentzen a better logician than himself.

Although this is a fact I have known well, I believe it is significant to communicate it here clearly. There are also examples of Gödel's views of Cohen, Kleene, Scott, Spector and others. I was already familiar with them, but I still enjoyed reading them.

There are many photos in this book. There is a photo of Gödel as a sweet youngest child in the family, some photos which Gödel himself displayed, and many others. I hope this book will be widely read in Japan as well as elsewhere.

Chapter 10

A Tribute to the Memory of Professor Gödel

Professor Gödel[1] died on January 14th, 1978 in Princeton. It is hardly necessary to restate his great achievements. Additionally, I cannot do them all justice in this brief note. I would like to pay a modest tribute to his memory with some personal recollections.

Professor Gödel was born on April 28th, 1906, in Brno, today in the Czech Republic. I believe that his father was the owner of a textile factory. Professor Gödel's health was fragile throughout his life, and his health was particularly poor in his childhood. His mother suffered from multiple sclerosis, and that made her concentrate on the education of her two sons.

Professor Gödel studied mathematics and physics at the Vienna University. His advisor was Hahn, famous for "the Hahn decomposition." Professor Gödel proved the Completeness Theorem in 1930, for which he was granted his Doctorate. This Completeness Theorem is the first genuine theorem in mathematical logic. Even now, it typifies theorems in mathematical logic and its proof is given at the beginning of courses in mathematical logic. It may be said that model theory and nonstandard analysis are all consequences of this theorem.

His Incompleteness Theorem was proved in 1931. This Incompleteness Theorem, according to Oppenheimer, "illuminated the role of limitations in human understanding in general." It should be seen as an extraordinary achievement of human knowledge, transcending mathematical logic and mathematics, sending out an eternal light. I shall forgo introducing this theorem. Instead, I hope that the reader will learn about it independently.

Professor Gödel served as a Privatdozent at the Vienna University from 1933 through 1938. In his last year there, he proved the consistency of the continuum hypothesis and the axiom of choice relative to ZF set theory.

---

[1] Gödel Sensei, in Japanese

This work is not only extraordinary by itself, but also it became a model for almost all progress in modern set theory as well as its starting point.

Professor Gödel visited the Institute for Advanced Study in Princeton several times between 1938 and 1940, and remained a regular member of the Institute from 1940 through 1953. During this period, some professors at the Institute who had great regard for him attempted to promote him to the rank of professor unsuccessfully with the objection of one professor. This made von Neumann, who revered him, deplore these circumstances and say: "How could I remain a professor when Gödel is not." At last, however, he became a professor in 1953.

He acquired American citizenship in 1948. The reason why he emigrated from Vienna to Princeton and acquired American citizenship was connected with the Nazis' rule of Germany and Austria, an unfortunate event of modern history.

In 1951 and 1952, honorary degrees were awarded to him respectively from Yale University and Harvard University. He was nominated as a member of United States' National Academy of Sciences, and in 1951 was awarded the Einstein prize. In 1976 he retired from his professorship at the Institute and became a professor emeritus.

I have listed only his three most famous works, but he left also other distinguished works relating to the decision problem, the theory of relativity and consistency proofs.

When I recall the life of Professor Gödel, what is most impressive is that he was filled with deep affection for scholarship, while constantly fighting against illness. He was always gentle to me and encouraged me pursuing my researches. Many of the logicians who were acquainted with him must have felt the same way. It seems to me that his weak constitution and his fight against illness were unusual. For this reason, he hardly went out anywhere except commuting between his home and the Institute. I suspect that the only occasion he attended a conference after coming to America was when he was awarded the Einstein prize. For this reason, he was often misunderstood to be a misanthrope or an eccentric, contrary to the fact, which is unfortunate.

The last occasion on which I discussed mathematics with Professor Gödel at length was about three years ago, when he invited me to the Institute for a week. During this time, he came to the Institute every day except for the day I gave a lecture, and talked with me in earnest for about two hours on each occasion. In fact I stayed at the Institute as a member three times, as I loved my occasional discussions with Professor Gödel. Previously our discussions lasted at most from thirty minutes to one hour. So, this was the first time that he had spent two hours with me. Furthermore, it happened every day. I was simultaneously glad and very much surprised. In fact, the subject of our talks then was a problem on the continuum. Recalling the discussions now, I feel like bowing to him with respect, and at the same time feel ashamed of my inability as his disciple to develop the subject in the direction of his ideas. He treated me in a manner that I remember fondly now. I felt sorry when we parted, and he too appeared to be so.

Not long after, the community of logicians began to learn that Professor Gödel was ill. We heard that he refused medical treatment. Many of us were concerned and tried to introduce him a good doctor and to persuade him to be hospitalized, but he refused all that. I heard that he had said, "I will die soon." I could understand his feeling on the basis of what I had heard from him previously. He had anticipated his death twice before, in 1946 and in 1970.

In 1946, when he thought his last hour was approaching, he communicated the proof of independence of the axiom of choice, which had been unpublished, to von Neumann. (Unfortunately, this proof is unpublished to date. When Cohen later proved the independence of the axiom of choice, this came up again. It is said that Professor Gödel was asked why he had not published the proof, and he answered: "I was afraid that it might lead logic in the wrong direction." I suspect that he was hesitant to publish his proof as it was not such a beautiful one as Cohen's.)

In 1970, an acquaintance of his introduced him to a famous doctor in New York and he received treatment. This doctor administered narcotics without informing him, which profoundly upset him, because he believed that narcotics damage the brain. This treatment was an extremely unpleasant experience and I imagine he was afraid more than anything else that it might be repeated.

The last occasion I conversed with him was last summer (1977). On the last day of the Conference in Durham, Tennenbaum made an international phone call to me. As I had gone to bed, his message: "Please make a phone call to Professor Gödel" reached me later. I did not understand what it was about, but in any case I phoned him upon returning to London. He only said "I am terribly ill." When I asked "Is there anything I can do for you?" he replied "Nothing." Days passed while I did not know what to do. I somehow hoped that he would recover again as had been the case in 1946 and 1970. But that was not to be.

According to Mrs. Gödel, Professor Gödel had been excited about his research to his last days. Whatever happened with his numerous unpublished works which had been written in that special shorthand? I do not know how to express his affection towards scholarship and his fight against disease. I believe that his great achievements and his deep insight will serve as a guide to us, his followers, to the proper direction of our subject for a long time. Beyond that, Professor Gödel's attitude towards scholarship will be an enduring model for us.

As I think of him I remember his gentle smile. In his last years, he was fond of reading books on philosophy and "The New York Times." Good bye, Professor Gödel.

Appendix A

On Gödel's Continuum Hypothesis

What I am going to describe here is on Gödel's[1]

"Some considerations leading to the probable conclusion that the true power of the continuum is $\aleph_2$."

As for the cardinality of the continuum, aside from the subject of this appendix, Gödel was seriously interested in the work of Hausdorff on Pantachie[2], and referred to one of Hausdorff's hypotheses concerning it as Hausdorff's continuum axiom.[3]

In relation to the work of Gödel along the line of Pantachie, I was asked a question by Solovay a few years ago. It was about a point in Gödel's notes in this area. I had found a misprint, and indicated it. From that contact, I imagine that Solovay is responsible for that portion of the edited Collected Works of Gödel, and that there are some shorthand notes of which I was not aware. I expect that they will clarify Gödel's thoughts on the subject.

What I shall take up here is independent of Gödel's work on Pantachie and related matters. It was dated later than his work on Pantachie, hence I believe that while he worked on this subject he thought it better than the work on Pantachie.

Gödel sent a five page memo to the Proceedings of the National Academy of Science. It was hand written, with nonsensical definitions and style, and it gives an impression of a doodle. The reason why this happened is that Gödel came under the care of a famous medical doctor

---

[1]This is a mathematical article, and was first printed September 10,1998.

[2]Paul du Bois-Remond, 1882, expressed the idea that the totality of orders of growth should be viewed as an expansion of the continuum (Walter Fletcher).

[3]The author refers the reader to "On Gödel's new axiom" in his book, written in Japanese, "Sugakukisoron no Sekai" (The World of the Foundations of Mathematics), Nippon Hyoron Sha,LTD.

in New York, who prescribed narcotics without informing him. Under the influence of narcotics, Gödel believed he had obtained a beautiful solution of a longstanding question, which he wrote out and submitted.

Unfortunately, there was an error in his reasoning, and the conclusion does not hold. What I shall explain here is my understanding of the picture of the continuum in his mind at the time.

In the article mentioned above, Gödel proposed an axiom which he believed was correct. Using that axiom he attempted to demonstrate that the cardinality of the continuum would be $\aleph_2$. Let me first define various notions and notations in order to state his theorem.

Let κ and λ be infinite cardinals where $\lambda \leq \kappa$. Define

$$^\kappa\lambda = \{h | h : \kappa \to \lambda\}.$$

For f, g such that $f, g \in {}^\kappa\lambda$, define

$$f \leq g \text{ by } \forall \alpha \in \kappa(f(\alpha) \leq g(\alpha))$$

and

$$f \leq_{ep} g \text{ by } \exists \alpha \in \kappa \forall \beta \in \kappa(\alpha \leq \beta \longrightarrow f(\beta) \leq g(\beta)).$$

(*ep* abbreviates 'end piece.')

Further, for $M \subset {}^\kappa\lambda$, define M^* by

$$M^* = \{f\lceil\alpha | f \in M \wedge \alpha < \kappa\}.$$

M is called major in $^\kappa\lambda$ if

$$M \subset {}^\kappa\lambda \text{ and } \forall f \in {}^\kappa\lambda \exists g \in M(f \leq g)$$

and M is called major by end piece in $^\kappa\lambda$ if

$$M \subset {}^\kappa\lambda \text{ and } \forall f \in {}^\kappa\lambda \exists g \in M(f \leq_{ep} g).$$

Denoting the cardinality of M by $|M|$, Gödel's new axiom takes the form as below.

Gödel's Square Axiom

For an arbitrary regular cardinal κ, there exists a subset of $^\kappa\kappa$, say M such that M is major by end piece in $^\kappa\kappa$ and satisfies the following two conditions.

1) M is well-ordered by $\leq_{ep}$, and its order type is κ^+.
2) $|M^*| = \kappa$.

Gödel often uses this axiom in an equivalent form:

$$\exists M((M \quad \text{is major in } {}^\kappa\kappa) \wedge |M| = \kappa^+ \wedge |M^*| = \kappa)$$

There is an article in which this axiom of Gödel's is studied:

G. Takeuti, Gödel numbers of product space, Higher Set Theory, edited by G.H. Müller and D.S.Scott, Lecture Notes in Math. ♯669, Springer 1978, pp. 461-471.

I will explain the content of this article in order to make use of its results.

First, for the aforesaid κ, λ, define $g(\kappa, \lambda)$ and $d(\kappa, \lambda)$ by the following equations.

$$g(\kappa, \lambda) = \min\{|F^*| | F \quad \text{is major in } {}^\kappa\lambda\}$$

$$d(\kappa, \lambda) = \min\{|M| | M \quad \text{is major in } {}^\kappa\lambda\}$$

Here $g(\kappa, \lambda)$ and $d(\kappa, \lambda)$ will be called respectively the First Gödel number and the Second Gödel number.

Theorem 1   Over ZFC, Gödel's axiom is equivalent to the following: for every regular cardinal κ

$$g(\kappa, \kappa) = \kappa \wedge d(\kappa, \kappa) = \kappa^+.$$

On these g and d, the following properties can be easily verified.

$$g(\omega, \omega) = \omega$$

$$d(\kappa, \lambda) = d(\kappa, cf(\lambda))$$

$$g(\kappa, \lambda) = g(\kappa, cf(\lambda))$$

$$\kappa^+ \leq d(\kappa, \lambda) \leq 2^\kappa$$

$$\kappa \leq g(\kappa, \lambda) \leq 2^{\overset{\kappa}{\smile}}$$

Here $2^{\overset{\kappa}{\smile}}$ is the limit of 2^μ for all μ below κ.

In the above mentioned article, as a corollary of Proposition 2, the following theorem is proven in ZFC.

Theorem 2   If $\lambda \leq \nu < \kappa$ and $\mu = d(\nu, \lambda)$, then the following inequalities hold.

   1) $g(\kappa, \mu) \leq g(\kappa, \lambda)$
   2) $d(\kappa, \mu) \leq d(\kappa, \lambda)$

As a corollary of this theorem, it obviously follows:

Corollary If $\lambda < \kappa$ and λ is a regular cardinal, then from Gödel's axiom, or more precisely from $d(\lambda, \lambda) = \lambda^+$, the inequalities below are derived.

   1) $g(\kappa, \lambda^+) \leq g(\kappa, \lambda)$
   2) $d(\kappa, \lambda^+) \leq d(\kappa, \lambda)$

By virtue of this corollary, it can be observed that $g(\kappa, \mu)$ is monotone nonincreasing with respect to μ as μ increases from λ to λ^+. Of course, this holds only locally when μ increases from λ to λ^+, and on the whole domain of μ the completely opposite phenomenon occurs as one can see from

$$d(\kappa, \lambda) = d(\kappa, cf(\lambda)), \quad g(\kappa, \lambda) = g(\kappa, cf(\lambda)).$$

Next, in the article mentioned above, the following theorem is proven in ZFC as a corollary of Theorem 1. [4]

Theorem 3   Under the condition $\kappa < 2^{\aleph_0}$ and $\omega < cf(\kappa)$, the following holds.

$$\kappa^+ \leq g(\kappa, \omega).$$

As a corollary of this theorem, the following holds.

Corollary The implication below follows from Gödel's axiom.

$$2^{\aleph_0} = \aleph_2 \rightarrow g(\aleph_1, \aleph_0) = g(\aleph_1, \aleph_1)^+.$$

Proof Substitute $\aleph_1$ for the κ in the theorem above to obtain

$$\aleph_2 \leq g(\aleph_1, \aleph_0).$$

On the other hand, we have

$$g(\aleph_1, \aleph_0) \leq 2^{\aleph_1} = 2^{\aleph_0} = \aleph_2.$$

---

[4]Both ω and $\aleph_0$ denote the least infinite ordinal. We comply with the author's usage of these letter.

Theorem 4 Under the assumption of Gödel's axiom, $2^{\aleph_0} = \aleph_2$ is equivalent to $H_1 \wedge H_2$, where H_1 and H_2 denote the following relations.

$H_1 : g(\aleph_1, \aleph_0) = g(\aleph_1, \aleph_1)^+$

H_2 : The least upper bound of $g(\aleph_1, \lambda)(\lambda \leq \aleph_1)$, that is, $2^{\aleph_1} = 2^{\aleph_0}$, is attained by $g(\aleph_1, \aleph_0)$.

Proof $2^{\aleph_0} = \aleph_2 \to H_1 \wedge H_2$ is obvious from the corollary of Theorem 3. Conversely, $H_1 \wedge H_2$ implies $2^{\aleph_0} = g(\aleph_1, \aleph_0) = \aleph_2$.

My view of what Gödel attempted was this. He first reduced the problem of $2^{\aleph_0}$ to $g(\aleph_\alpha, \aleph_\beta)$ and $d(\aleph_\alpha, \aleph_\beta)$, which he was able to think of more concretely. Dealing with these objects, he proposed a promising axiom, with which he would resolve the problem.

I would like to propose the axiom below, as an extension of Gödel's idea in this sense.

First, consider $g(\aleph_k, \aleph_i)$ where $i \leq k$ and i and k are both natural numbers. In such a case, phenomena which are similar to those of Gödel's cases are expected to occur.

g-regularity Axiom

For i and k natural numbers, where $i < k$, holds

$$g(\aleph_k, \aleph_i) = g(\aleph_k, \aleph_{i+1})^+.$$

This axiom means that change from $g(\aleph_k, \aleph_i)$ to $g(\aleph_k, \aleph_{i+1})$ is well-regulated.

Assuming this axiom, it immediately follows

$$g(\aleph_k, \aleph_0) = \aleph_{2k}$$

and, using further $2^{\aleph_k} = 2^{\aleph_{k+1}} \geq g(\aleph_{k+1}, \aleph_0)$, one can immediately obtain

$$2^{\aleph_k} \geq \aleph_{2(k+1)} \quad (k < \omega).$$

g-fullness Axiom

For every i, where $0 < i < \omega$, the following holds.

$$g(\aleph_i, \aleph_0) = 2^{\aleph_{i-1}}$$

This axiom tells us that, since $g(\aleph_i, \aleph_0) \leq 2^{\aleph_i} = 2^{\aleph_{i-1}}$, $g(\aleph_i, \aleph_0)$ is assumed to be the maximum possible value.

If one adds this axiom, then

$$2^{\aleph_k} = \aleph_{2(k+1)} \qquad (k < \omega)$$

necessarily holds, and therefore $2^{\aleph_k}$ is completely determined for $k < \omega$.

I would subsequently like to state an extension of Gödel's idea rather freely. Let us first add the following axiom H to Gödel's axioms.

H: For a regular $\aleph_\alpha$, the least upper bound of $g(\aleph_\alpha, \lambda)$ $(\lambda \leq \aleph_\alpha)$, that is $2^{\aleph_\alpha}$, can be attained by $g(\aleph_\alpha, \lambda)$ for a certain λ.

Subsequently κ and $\aleph_\alpha$ are all assumed to be smaller than the first weakly inaccessible cardinal.

Under this assumption, the axiom H and the equation

$$2^{\aleph_\alpha} = 2^{\aleph_{\alpha+1}} = g(\aleph_{\alpha+1}, \aleph_0)$$

are equivalent by virtue of the corollary of Theorem 2 and the equation $g(\kappa, \lambda) = g(\kappa, cf(\lambda))$. It is therefore sufficient to evaluate $g(\aleph_{\alpha+1}, \aleph_0)$ in order to evaluate $2^{\aleph_\alpha}$.

We subsequently employ the computation procedure below as a set of postulates.

It is a procedure to evaluate $g(\aleph_{\alpha+1}, \aleph_0)$ starting with $g(\aleph_{\alpha+1}, \aleph_{\alpha+1})$ and gradually decreasing the λ in $g(\aleph_{\alpha+1}, \lambda)$. Needless to say that we consider below the first weakly inaccessible cardinal.

1. $g(\aleph_{\alpha+1}, \aleph_{\alpha+1})$ is admissible and $g(\aleph_{\alpha+1}, \aleph_{\alpha+1}) = \aleph_{\alpha+1}$ (Gödel's axiom).

2. If $g(\aleph_{\alpha+1}, \aleph_{\alpha+1})$ is admissible and $g(\aleph_{\alpha+1}, \aleph_{\beta+1}) = \mu$, then $g(\aleph_{\alpha+1}, \aleph_\beta)$ is also admissible and $g(\aleph_{\alpha+1}, \aleph_\beta) = \mu^+$.

3. If $g(\aleph_{\alpha+1}, \aleph_\beta)$ is admissible and $g(\aleph_{\alpha+1}, \aleph_\beta) = \mu$, then $g(\aleph_{\alpha+1}, cf(\aleph_\beta))$ is also admissible and $g(\aleph_{\alpha+1}, cf(\aleph_\beta)) = \mu$.

It should be obvious that $g(\aleph_{\alpha+1}, \aleph_0)$, and hence $2^{\aleph_0}$, can be evaluated with the postulates above. We will give here two examples.

1. Evaluation of $2^{\aleph_{\omega+1}}$

We can obtain the conclusion successively as follows.

$$g(\aleph_{\omega+2}, \aleph_{\omega+2}) = \aleph_{\omega+2}$$

$$g(\aleph_{\omega+2}, \aleph_\omega) = \aleph_{\omega+4}$$

$$g(\aleph_{\omega+2}, \aleph_0) = \aleph_{\omega+4}$$

$$2^{\aleph_{\omega+1}} = \aleph_{\omega+4}$$

2. Evaluation of $2^{\aleph_{\omega_1+2}}$

We can obtain the conclusion successively as follows.

$$g(\aleph_{\omega_1+3}, \aleph_{\omega_1+3}) = \aleph_{\omega_1+3}$$

$$g(\aleph_{\omega_1+3}, \aleph_{\omega_1}) = \aleph_{\omega_1+6}$$

$$g(\aleph_{\omega_1+3}, \aleph_{\omega_1}) = \aleph_{\omega_1+6}$$

$$g(\aleph_{\omega_1+3}, \aleph_0) = \aleph_{\omega_1+7}$$

$$2^{\aleph_{\omega_1+2}} = \aleph_{\omega_1+7}$$

What should we do with evaluation of $g(\aleph_{\alpha+1}, \aleph_\beta)$ when it is not admissible? Obviously it will be made possible if we add the following two rules.

1. If $g(\aleph_{\alpha+1}, \aleph_\beta) = \mu^+$, then $g(\aleph_{\alpha+1}, \aleph_{\beta+1}) = \mu$.
2. If $g(\aleph_{\alpha+1}, \aleph_\beta) = \mu$ and μ is a limit cardinal, then $g(\aleph_{\alpha+1}, \aleph_{\beta+1}) = \mu$.

Under Gödel's influence I once gave some thought to his approach to the continuum hypothesis. After twenty years of neglect I have here improved on my old thoughts a little. I would be pleased if this attracted someone's interest.

Appendix B

Birth of Second Order Proof Theory by the Fundamental Conjecture on GLC

It was during my student days in 1946 at the Tokyo Imperial University (now the University of Tokyo) when I began to study the foundations of mathematics. At that time, at the Tokyo Imperial University, courses of study were divided from the start,[5] and I enrolled as a student in the Mathematics Department. The undergraduate program lasted for three years, and these years were called respectively the first year, the middle year and the last year. In the last year of mathematical studies, each student was required to select an advisor, choose a field of specialization and to give talks at seminars under the guidance of his advisor.

My advisor was Professor Shokichi Iyanaga, and my area was analytic set theory. I followed the notes of Survey Lectures by Luzin (Lusin in the earlier spelling) and, having worked through them, I undertook Luzin's three most difficult problems. I thought of them day and night. It was in the postwar period and Japan suffered from a food shortage. Almost every night, I would awaken with a feeling of hunger while these problems coursed vividly through my mind. After two months of struggle, I concluded that I could not solve these problems. It seemed it was in part because of the terrible privations of postwar life. Soon, however, I made some lucky discoveries: it was not the result of my limitations that I could not solve these problems; they are unsolvable! For various reasons, which I no longer remember, my conviction grew. I explained my ideas to Professor Iyanaga earnestly, and asked him how I could possibly approach showing that these problems were unsolvable. Notwithstanding that he was quite busy, he carefully listened to me, and suggested to this end that I might study the foundations of mathematics, and he introduced J. Herbrand's paper to me. It was not available in the department's library, so he then suggested that

---

[5]Now a student decides the major during the second year.

I read G. Gentzen's articles. That is why I read Gentzen, and that is how my mathematical career began.

Gentzen's works belong to the foundations of mathematics, specifically to proof theory. The foundations of mathematics originated in the discovery of contradictions which arose in naive set theory. Since the notion of sets is fundamental for modern mathematics, contradictions arising in set theory are a serious matter. In the current day, axiomatic set theory is fully accepted and it is generally acknowledged that modern mathematics can be carried out in the framework of axiomatic set theory. No contradiction has arisen in axiomatic set theory, and a sense of security that no contradiction will arise in it in the future is supported by intuitive consensus. Under the current secure circumstances one cannot imagine the sense of crisis of that earlier time. Was it an overreaction? Or, are we, present day mathematicians, too complacent and mechanically involved with our achievements? If I may add one word to the latter, I must assert that the contemporary perspective of the universe of set theory is shallow. This phenomenon is seen clearly in the lack of effort to reach for a new axiom of set theory.

It was foremost in Hilbert's mind to rescue mathematics from contradictions in set theory.

To this end, he proposed the program, now called Hilbert's Program, described below.

(1) Formalize mathematics and treat it as a formal system.
    For this reason, Hilbert's standpoint is called formalism.
(2) In this formalized system, a proposition is expressed with a string which is composed of some special symbols which stand for basic propositions, variables and logical notions respectively. Mathematical inferences and proofs are all described as concrete rules which apply to these strings, which are clusters of symbols as above.
(3) Only the arguments which are applied to the formalized proofs that are concretely considered as above and which are simple and reliable can be admitted. Namely, the strings which are relevant here are concrete strings consisting of finitely many symbols. This standpoint is called the finite standpoint.
(4) Complete the foundational underpinnings of the formalized mathematical system by proving that no contradiction can be derived in this formal system. That is, demonstrate that the system is consistent, within the finite standpoint.

In this Program of Hilbert's, "the arguments which were applied to the formalized proofs" served to advance proof theory, the study of formal proofs. Since the formalized system can be identified with logic itself, it would certainly be a serious mistake if all of our considerations of this research were restricted to Hilbert's Program alone. Nevertheless Hilbert's Program served as a powerful engine moving proof theory forward. Prior to Gentzen's discoveries, there had been K. Gödel's famous Incompleteness Theorem, and it influenced Gentzen. I will explain this later. Here I would like to introduce Gentzen's works. Since the purpose is an overview, I will confine myself to the logical symbols ¬ (negation), ∧ (and) and ∀ (for all). Three other logical symbols, ∨ (or), ⊃ (implies) and ∃ (exists), can be expressed by combinations of these three, ¬, ∧, ∀.

Gentzen introduced the logical system LK. [6] LK uses not only formulas but also sequents. A sequent is an expression of the form

$$A_1, \cdots, A_m \longrightarrow B_1, \cdots, B_n,$$

where $A_1, \cdots, A_m, B_1, \cdots, B_n$ are formulas, and $m \geq 0, n \geq 0$. Notice that $m = 0$ or $n = 0$ are allowed. The meaning of the sequent as above is the following.

Under the assumption of $A_1, \cdots, A_m$, at least one of $B_1, \cdots, B_n$ holds.

Namely, it means

$$A_1 \wedge \cdots \wedge A_m \longrightarrow B_1 \vee \cdots \vee B_n.$$

Here when $m = 0$, $\longrightarrow B_1, \cdots, B_n$ means "at least one of $B_1, \cdots, B_n$ holds." When $n = 0$, $A_1, \cdots, A_m \longrightarrow$ means "a contradiction is derived from the assumption $A_1, \cdots, A_m$."

In the extreme case of $m = n = 0$, the sequent assumes the form $\longrightarrow$, and it means "a contradiction occurs without assumptions," that is, the sequent is a contradiction itself. $\longrightarrow$ therefore certainly is not derived in LK. [7]

A proof in the system LK starts with sequents of the form $D \longrightarrow D$ and inferences cumulatively transform the sequents. Since inferences are applied to sequents of the form $A_1, \cdots, A_m \longrightarrow B_1, \cdots, B_n$, very often finite sequences of formulas are dealt with. For this reason, it is convenient

---

[6] A system of predicate calculus.
[7] LK is known to be consistent.

to introduce Greek upper case letters such as $\Gamma, \Delta, \cdots$ as symbols expressing finite sequences of formulas. Here a finite sequence can stand for the empty sequence. For example, when Γ and Δ are both empty, $\Gamma \longrightarrow \Delta$ denotes $\longrightarrow$.

There are two kinds of inference rules in LK. One concerns the logical symbols, while the other concerns the structure of a sequent. The latter is necessary because we employ the sequents. They involve the structures of sequents, and hence they do not have logical meaning. To be precise, I will write down the inference rules one by one.

$$\frac{\Gamma, C, D, \Pi \longrightarrow \Delta}{\Gamma, D, C, \Pi \longrightarrow \Delta} \; (left \; exchange) \qquad \frac{\Gamma \longrightarrow \Delta, C, D, \Lambda}{\Gamma \longrightarrow \Delta, D, C, \Lambda} \; (right \; exchange)$$

Each sequent above a horizontal line is called the upper sequent of the inference, and each sequent below a horizontal line is called the lower sequent of the inference. An inference is represented in a way that from the upper sequent the lower sequent is inferred.

In the left inference rule above, the upper sequent is of the form $\Gamma, C, D, \Pi \longrightarrow \Delta$ and the lower sequent is of the form $\Gamma, D, C, \Pi \longrightarrow \Delta$. This inference rule expresses that, in the assumption part [8] of a sequent, the order of formulas can be exchanged. The right inference rule above expresses that "in the conclusion part [9] of a sequent, the order of formulas can be exchanged."

$$\frac{\Gamma \longrightarrow \Delta}{D, \Gamma \longrightarrow \Delta} \; (left \; weakening) \qquad \frac{\Gamma \longrightarrow \Delta}{\Gamma \longrightarrow \Delta, D} \; (right \; weakening)$$

The meaning of the left inference rule is that, if $\Gamma \longrightarrow \Delta$ holds, then it will hold when an extra assumption is added to the assumption Γ. It is obvious that $D, \Gamma \longrightarrow \Delta$ is logically weaker than $\Gamma \longrightarrow \Delta$, and hence the above holds for an arbitrary formula D. Similarly, the right inference rule asserts that, if $\Gamma \longrightarrow \Delta$ holds, then the sequent $\Gamma \longrightarrow \Delta, D$ with an extra conclusion D also holds.

$$\frac{D, D, \Gamma \longrightarrow \Delta}{D, \Gamma \longrightarrow \Delta} \; (left \; contraction) \qquad \frac{\Gamma \longrightarrow \Delta, D, D}{\Gamma \longrightarrow \Delta, D} \; (right \; contraction)$$

The left inference rule asserts that having two identical assumptions and having one assumption amount to the same. The right inference rule asserts

---

[8] the sequence of formulas in the left side of a sequent

[9] the sequence of formulas in the right side of a sequent

that having two identical conclusions and having one of them amount to the same. I believe it is obvious that those are correct inference rules.

Now, the first important inference rule of LK is the following "cut."

$$\frac{\Gamma \longrightarrow \Delta, D \quad D, \Pi \longrightarrow \Lambda}{\Gamma, \Pi \longrightarrow \Delta, \Lambda} \; (cut)$$

In this inference rule, there are two upper sequents, $\Gamma \longrightarrow \Delta, D$ and $D, \Pi \longrightarrow \Lambda$. Of course, this inference rule asserts that if two upper sequents are correct, then the lower sequent $\Gamma, \Pi \longrightarrow \Delta, \Lambda$ is also correct.

This inference rule is a generalization of the syllogism below.

$$\frac{A \longrightarrow B \quad B \longrightarrow C}{A \longrightarrow C}$$

Correctness of the cut can be shown the same way as the correctness of the syllogism.

Other than the inference rules above, there are inference rules concerning logical symbols.

First, on negation:

$$\frac{\Gamma \longrightarrow \Delta, D}{\neg D, \Gamma \longrightarrow \Delta} \; (left \; \neg) \qquad \frac{D, \Gamma \longrightarrow \Delta}{\Gamma \longrightarrow \Delta, \neg D} \; (right \; \neg)$$

The left rule of inference can be obtained from the law of contradiction $D, \neg D \longrightarrow$ by applying a cut [10]:

$$\frac{\Gamma \longrightarrow \Delta, D \quad D, \neg D \longrightarrow}{\neg D, \Gamma \longrightarrow \Delta} \; (cut)$$

The converse is also true; the law of contradiction can be derived by

$$\frac{D \longrightarrow D}{\neg D, D \longrightarrow} \; (left \; \neg)$$

with $\Gamma \equiv D$ and Δ being empty. That is to say, the rule "left $\neg$" is the law of contradiction expressed in the form of an inference. Similarly, the right inference rule is obtained from the law of the excluded middle $\longrightarrow D, \neg D$ by applying a cut:

$$\frac{\longrightarrow D, \neg D \quad D, \Gamma \longrightarrow \Delta}{\Gamma \longrightarrow \Delta, \neg D} \; (cut)$$

---

[10]and some exchanges

and the law of the excluded middle can be obtained by an application of "right ¬." That is, it is the law of the excluded middle expressed in the form of an inference. I will explain later what wonderful properties follow from writing some laws in the form of inference rules.

The inference rule of $\wedge$ is the following.

$$\frac{A, \Gamma \longrightarrow \Delta}{A \wedge B, \Gamma \longrightarrow \Delta} \ (left \ \wedge) \qquad \frac{B, \Gamma \longrightarrow \Delta}{A \wedge B, \Gamma \longrightarrow \Delta} \ (left \ \wedge)$$

In these rules of inference, $A \wedge B$ is introduced in the left side of a sequent. Both are derived from respectively $A \wedge B \longrightarrow A$ and $A \wedge B \longrightarrow B$, which are valid from the meaning of $\wedge$, by applications of the cut rule, similarly to the case of ¬.

The rule of inference in which $A \wedge B$ is introduced in the right side of a sequent is the following.

$$\frac{\Gamma \longrightarrow \Delta, A \quad \Pi \longrightarrow \Lambda, B}{\Gamma, \Pi \longrightarrow \Delta, \Lambda, A \wedge B} \ (right \ \wedge)$$

This too is derived from $A, B \longrightarrow A \wedge B$, which is valid from the meaning of $\wedge$, by applications of the cut rule twice:

$$\frac{\Gamma \longrightarrow \Delta, A \quad \dfrac{\Pi \longrightarrow \Lambda, B \quad A, B \longrightarrow A \wedge B}{A, \Pi \longrightarrow \Lambda, A \wedge B} \ (cut)}{\Gamma, \Pi \longrightarrow \Delta, \Lambda, A \wedge B} \ (cut)$$

Finally, the inference rules of $\forall$ are the following.

$$\frac{F(t), \Gamma \longrightarrow \Delta}{\forall x F(x), \Gamma \longrightarrow \Delta} \ (left \ \forall) \qquad \frac{\Gamma \longrightarrow \Delta, F(a)}{\Gamma \longrightarrow \Delta, \forall x F(x)} \ (right \ \forall)$$

(t is an arbitrary term, and a is a free variable not occurring in the lower sequent.)

The left rule of inference is an expression of the sequent $\forall x F(x) \longrightarrow F(t)$.

In the right rule of inference above, the condition, known as the eigen-variable condition, that a does not occur in the lower sequent is important. a in this context is called an eigenvariable of the inference. This inference is a generalization of

$$\frac{\longrightarrow F(a)}{\longrightarrow \forall x F(x)}$$

when a satisfies the eigenvariable condition. This expresses the meaning of $\forall$, that is, "if $F(a)$ holds for an arbitrary a, then $\forall x F(x)$ holds." The rule "right $\forall$" can be justified by means of this special case.

This exhausts the inference rules. Starting with sequents of the form $D \longrightarrow D$ and applying these rules of inference, all the correct sequents of the predicate calculus can be proved in LK. Now, the fundamental theorem below tells us why the system LK, in which usual logical axioms are expressed in terms of inference rules, is so wonderful.

Fundamental Theorem of LK

If $\Gamma \longrightarrow \Delta$ is provable in LK, then $\Gamma \longrightarrow \Delta$ can be proved in LK without the cut rule.

It may not be clear why this theorem is so wonderful at a glance. However, look at any rule of inference other than the cut rule, "left $\forall$" for example.

$$\frac{F(t), \Gamma \longrightarrow \Delta}{\forall x F(x), \Gamma \longrightarrow \Delta}$$

A formula in the upper sequent is either the same as or simpler than the corresponding formula in the lower sequent. A formula either in Γ or Δ is shared with the upper and the lower sequent. We only have to compare $F(t)$ and $\forall x F(x)$, and certainly $\forall x F(x)$ has one extra logical symbol compared with $F(t)$. That is, the lower sequent is logically more complex than the upper sequent. On the other hand, the cut rule violates this condition.

Suppose in

$$\frac{\Gamma \longrightarrow \Delta, D \quad D, \Pi \longrightarrow \Lambda}{\Gamma, \Pi \longrightarrow \Delta, \Lambda} \ (cut)$$

D is logically complex, while all formulas in $\Gamma, \Pi, \Delta, \Lambda$ are logically simple. Then this cut permits the inference of a logically simple sequent from two logically complex upper sequents. The fundamental theorem therefore assures us that a theorem of the predicate calculus can be proved step by step from logically simpler sequents. For example, consider a situation in which an elegant and simple theorem-let has been proved from a complex big theorem. The fundamental theorem asserts that in such a case the elegant theorem-let can be proved in a much simpler way. The fundamental theorem is quite powerful, and it has many useful consequences.

Here let me take up the simplest possible example. Namely, I will show that no contradiction arises in LK, or that the sequent $\longrightarrow$ cannot be proved in LK.

Suppose contrary that $\longrightarrow$ were proved in LK. Then by the fundamental theorem there is a cut-free proof of $\longrightarrow$ in LK. It starts with a sequent of the form $D \longrightarrow D$. The last sequent is $\longrightarrow$, in which no logical symbol occurs. Let us try to find the first inference in the proof at which a sequent without any formula occurs. With respect to any inference except a cut, a formula occurring in the upper sequent occurs in the lower sequent either by itself or as a part of another formula. So, it is impossible to find such a sequent in the proof, which implies that the sequent $\longrightarrow$ could not have been proved.

Now, the most famous achievement of Gentzen is the consistency proof of the system of number theory. This was an epoch making event in the history of Hilbert's Program. As I view this consistency proof of the system of number theory, the crucial part is in its essence the fundamental theorem of LK as explained above. Stating more clearly, the proof that no contradiction is derived in the system of number theory is based on the idea of the demonstration that $\longrightarrow$ is not derived in LK. Let me explain. In formalizing number theory, one adds simple Peano axioms such as $\longrightarrow a + b = b + a$ besides the axioms of the form $D \longrightarrow D$; these are starting sequents in a proof of the system. The characteristics of these new mathematical axioms are that they are sequents without logical symbols. With these axioms added, the system of number theory is obtained from LK by adding a rule of inference called Ind:

$$\frac{A(a), \Gamma \longrightarrow \Delta, A(a+1)}{A(0), \Gamma \longrightarrow \Delta, A(t)} \ (Ind)$$

Here a is assumed to satisfy the eigenvariable condition. That is to say, a does not occur in the lower sequent.

This inference rule is an expression of the axiom of mathematical induction

$$A(0) \wedge \forall x(A(x) \supset A(x+1)) \longrightarrow \forall x A(x).$$

Having axiomatized number theory in this way, the consistency proof of the system of number theory is an extension of the demonstration of the following fundamental theorem. It can be explained as follows.

Let t represent a natural number $n = 1 + 1 + \cdots + 1$. Then the induction inference

$$\frac{A(a), \Gamma \longrightarrow \Delta, A(a+1)}{A(0), \Gamma \longrightarrow \Delta, A(n)}$$

can be replaced with the following $n - 1$ many cuts.

$$\frac{\dfrac{A(0), \Gamma \longrightarrow \Delta, A(1) \quad A(1), \Gamma \longrightarrow \Delta, A(2)}{A(0), \Gamma \longrightarrow \Delta, A(2)} \quad A(2), \Gamma \longrightarrow \Delta, A(3)}{A(0), \Gamma \longrightarrow \Delta, A(3)}$$

$$\vdots$$

$$\frac{A(0), \Gamma \longrightarrow \Delta, A(n-1) \qquad A(n-1), \Gamma \longrightarrow \Delta, A(n)}{A(0), \Gamma \longrightarrow \Delta, A(n)}$$

Here $A(i), \Gamma \longrightarrow \Delta, A(i+1)$ is obtained from $A(a), \Gamma \longrightarrow \Delta, A(a+1)$ by substituting i for a.

Also, the lower sequent of the cut in

$$\frac{A(0), \Gamma \longrightarrow \Delta, A(1) \quad A(1), \Gamma \longrightarrow \Delta, A(2)}{A(0), \Gamma \longrightarrow \Delta, A(2)}$$

is, more precisely, obtained from $A(0), \Gamma, \Gamma \longrightarrow \Delta, \Delta, A(2)$ by some contractions.

This suggests that, if a contradiction, or $\longrightarrow$, is derived in the system of number theory, then one can replace the Ind rule by cuts one by one, and later can eliminate cuts. Then similarly to the case of LK, one can show that in fact the sequent $\longrightarrow$ cannot be derived. Namely, Gentzen's consistency proof of the system of number theory can be constructed as an extension of the proof of the fundamental theorem of LK, a fundamental theorem on logic. That is why it is beautiful and it convinces us of correctness of number theory.

Now as I mentioned earlier, there had been a recent major discovery in the foundations of mathematics, Gödel's Incompleteness Theorem, before Gentzen's work. Gödel's Incompleteness Theorem is composed of the following two theorems.

1. Any consistent axiomatic system which contains number theory necessarily has an undecidable proposition. Furthermore, it is a proposition about number theory and is expressed in terms of symbols of number theory. In other words, there is a number theoretic proposition about which it cannot be proved within the system whether it holds or not.

2.   Within a consistent axiomatic system, its consistency (a proposition which is a formalization of the consistency and is expressed in terms of number theoretic language) cannot be derived.

The first theorem was recognized as sensational, since at that time almost every one had believed that any number theoretic proposition would be decided in set theory. However, it was the second theorem that delivered the fatal blow to Hilbert's Program. Since Hilbert's finite standpoint was naive and elementary, it was hard to believe that there could be an inference that could not be formalized in a formal system of number theory.

In that case, where is an argument in Gentzen's consistency proof of the formal system of number theory that cannot be formalized in number theory? First, the proof of the fundamental theorem can be executed in a completely elementary and finite standpoint, and so can be formalized in number theory. In Gentzen's proof, as I wrote earlier, each mathematical induction is reduced to many cuts. Then these cuts are eliminated. On the whole, many mathematical inductions are replaced with many cuts, and then those cuts are eliminated. This process is repeatedly executed. As we observe this process, we see that it reduces to the accessibility problem of transfinite ordinal numbers up to ε_0. Here ε_0 is an ordinal which is the limit of $\omega_1, \omega_2, \omega_3, \cdots$.[11] An ordinal number α is said to be accessible if any decreasing sequence of ordinal numbers below α, say $\alpha_1 > \alpha_2 > \alpha_3 > \cdots$, halts after a finite number of steps. The fact that ε_0 is accessible can roughly be shown as below.

- ω is accessible. Given any α such that $\omega > \alpha$, α is a natural number n. So, any decreasing sequence below α halts at most within n steps.
- A proof of accessibility of ω^ω can be reduced to accessibility of ω^n for all natural number n. This is shown, very roughly, in a manner similar to the proof of 1 above. Accessibility of ω^n can be reduced to that of ω^{n-1}. Repeating the same argument n times, accessibility can be reduced to that of ω.
- Similarly, accessibility of ω^{ω^ω} is reduced to accessibility of ω^ω. Repeating this process, accessibility of ε_0 can be reduced to accessibility of all ω_n.

There are various arguments about in what sense Gentzen's consistency proof departs from Hilbert's original naive finite standpoint, and in what sense it still inherits the spirit of Hilbert's finite standpoint and can be

---

[11]$\omega_1 = \omega, \omega_{n+1} = \omega^{\omega_n}$

regarded as a kind of finite standpoint. Opinions diverge depending upon whether one advocates the finite standpoint or opposes it.

When I began to study the foundations of mathematics, it had advanced as far as I have already indicated. With the difficulties of the postwar period, we could obtain publications only as late as 1938. However, the advancement of scholarship in the area had temporarily halted worldwide during the Second World War. The Hilbert School was especially in decline due to Gödel's Incompleteness Theorem and the policies of the Nazis in Germany. I can claim confidently that there was no general urge to study the consistency problem, especially to study the notion of impredicative sets proof-theoretically. However, as my objective was a proof theoretic study of the notion of sets from the beginning, I plunged into such a study without misgivings. Of course I knew nothing of what was happening in the world, and so I simply concentrated on what I wanted to do.

The notion of sets is a translation of propositional variables. For example, consider a set of natural numbers A and the proposition that a natural number n is an element of A, expressed as $n \in A$. It is the same as taking a unary propositional variable α over natural numbers and considering $\alpha(n)$:

$$n \in A \longleftrightarrow \alpha(n).$$

Namely, considering the notion of sets over a specific domain is equivalent to considering the second order predicate logic. That is, one introduces second order predicate variables $\alpha, \beta, \cdots, \phi, \psi, \cdots$, and, for each term t, regards an expression like $\alpha(t)$ as a formula, and then admits a second order $\forall \phi$ similarly to the first order $\forall x$.

Corresponding to the notion of a set $\{x|A(x)\}$, define a form $\{x\}A(x)$. The expression $\{x\}A(x)$ does not occur in forming formulas of second order predicate logic, but as I will explain later we use it in defining rules of inference on formulas. Having prepared in this way, the second order predicate logic G^1LC, which is an extension of LK, is obtained from LK by augmenting the inference rules below.

$$\frac{F(\{x\}A(x)), \Gamma \longrightarrow \Delta}{\forall \phi F(\phi), \Gamma \longrightarrow \Delta} \ (left \ second \ order \ \forall)$$

$$\frac{\Gamma \longrightarrow \Delta, F(\alpha)}{\Gamma \longrightarrow \Delta, \forall \phi F(\phi)} \ (right \ second \ order \ \forall)$$

Here $A(\alpha)$ denotes an arbitrary formula. α satisfies the eigenvariable condition that α does not occur in the lower sequent.

Let me note one thing here. Let $F(\{x\}A(x))$ be the formula

$$A(0) \wedge \forall x (A(x) \supset A(x+1)) \supset A(t).$$

Then $\forall \phi F(\phi)$ is

$$\forall \phi (\phi(0) \wedge \forall x (\phi(x) \supset \phi(x+1)) \supset \phi(t)).$$

In fact, I had originally defined a system of logic called GLC (generalized logical calculus), which contains predicate calculi of all finite orders, and in particular studied G^1LC, a second order subsystem of GLC. Since most of subsequent works of proof theory concerns second order subsystems, I will confine myself to the story about G^1LC in most cases. It was in such a setting that I proposed my fundamental conjecture on GLC.

The fundamental conjecture on GLC: If $\Gamma \longrightarrow \Delta$ is provable in GCL, then $\Gamma \longrightarrow \Delta$ is provable in GLC without applications of the cut rule.

At the same time that I proposed the fundamental conjecture, I demonstrated that the consistency of analysis can be derived from the fundamental conjecture in the elementary, finite standpoint in the sense of Hilbert. In Gentzen style proof theory, it is important to reduce the consistency problem to the fundamental logical problem. My fundamental conjecture was one such attempt. The fundamental conjecture on GLC has also other applications. However, the fundamental conjecture on GLC has a feature that is entirely different from Gentzen's fundamental theorem. From Gentzen's fundamental theorem, an important result follows, that is, a sequent is derived in the system from simple sequents by accumulating more complex ones. This does not follow from the fundamental conjecture of GLC. The cause of this failure is in the rule of 'left second order $\forall$':

$$\frac{F(\{x\}A(x)), \Gamma \longrightarrow \Delta}{\forall \phi F(\phi), \Gamma \longrightarrow \Delta}$$

In this inference, if $A(x)$ is a complex formula, then $F(\{x\}A(x))$ is far more complex than $\forall \phi F(\phi)$, and hence the principle that the upper sequent is simpler than the lower sequent does not hold.

The rule of inference above is in fact an expression of the comprehension axiom[12] in the form of an inference. The case where $F(\{x\}A(x))$ is far more

---

[12]For a formula $A(x)$, the comprehension axiom is an axiom of the form $\exists \phi \forall x (\phi(x) \leftrightarrow A(x))$, where $\exists x$ expresses 'there exists x.'

complex than $\forall\phi F(\phi)$ is called an impredicative case, which is related to a paradox in naive set theory. It is a case of the so-called vicious circle, and is a central difficulty in defining the notion of sets. This also causes difficulty in analyzing higher order predicate logics such as GLC, and hence causes a problem in attempts to prove the fundamental conjecture.

Having proposed the fundamental conjecture, I concentrated on its proof and spent several years in an anguished struggle trying to resolve the problem day and night. Here I would like to talk about memories of my late friend Shoji Maehara. It was on the way home from a conference when I first met Maehara. We talked about the foundations of mathematics a bit, and as a result he came to visit me at my house. On the occasion of our second meeting, I told him about my fundamental conjecture and the way I was desperately struggling with it. Maehara told me: "If I were you, I would not try to resolve the whole fundamental conjecture. It would be interesting enough if your problem were resolved for a subsystem which includes, for example, number theory." This was a great revelation to me. I was then able to write down various results which I could derive from my desperate struggles. I thus began to publish partial results on the fundamental conjecture. Maehara himself did not work on the fundamental conjecture, but his friendship as well as his support was a great help to me. Although I hesitate to write about personal matters here, if I may be allowed to list the names of other people who gave assistance to me, they are Takakazu Shimauchi, Akiko Kino and Mariko Yasugi.

With all that, the subsystems for which I have been able to prove the fundamental conjecture are the system with Π_1^1 comprehension axiom and a slightly stronger system, that is, the one with Π_1^1 comprehension axiom together with inductive definitions. Incidentally, according to research by Harvey Friedman, the system with the Δ_2^1 comprehension axiom can be reduced to the system with the Π_1^1 comprehension axiom and inductive definitions. Since Friedman showed this fact by applying nonstandard number theory, his method is remote from Hilbert's Program. Mariko Yasugi and I tried to resolve the fundamental conjecture for the system with the Δ_2^1 comprehension axiom within our extended version of the finite standpoint. Ultimately, our success was limited to the system with provably Δ_2^1 comprehension axiom. This was my last successful result in this area.

My fundamental conjecture itself has been resolved in a sense by Motoo Takahashi and Dag Prawitz independently. However, their proofs rely on set theory, and so it cannot be regarded as an execution of Hilbert's Program. On the other hand, according to our approach, we can see what sort of reductions are applied and how cuts can be gradually eliminated. So, our approach supplies us with information on the properties of the system in concern, the proof theoretic strength of the system in particular, which implies varieties of applications. With a set theoretic approach, however, I do not believe we can get such information, and hence it is not useful for certain problems. Following our research, it was mainly the Schütte School who developed proof theory. As a consequence, the fundamental conjecture has been resolved for the case of the system with the Π_2^1 comprehension axiom independently by Michael Rathjen and Toshiyasu Arai. This is a definite advancement from our result. It is a blessing result and I expect further developments along their line.

In this article, I have given an elementary exposition on second order proof theory from my personal standpoint. Looking back over my past, having gone to America, I lived away from friends and collaborators, and, since proof theory was not so popular in America, I became isolated. My collaborators, Akiko Kino and Mariko Yasugi, left me one by one. My research thus became solitary. I felt that, even if I published partial results, nobody would be interested in them, and so I quit publishing and my research interest shifted to another field. I wish the Schütte School had caught up with proof theory a little earlier.

Nevertheless, I am confident that, having proposed the fundamental conjecture and having resolved it for some subsystems, I initiated the study of second order proof theory.

If I may add another digression, Gentzen had already died in a prison in Prague when I entered this area. Of course I was unaware of that at the time. When I had written the article on the fundamental conjecture, I believed that in Germany, research in this area was prospering. My article on the fundamental conjecture was reviewed by a German mathematical logician. In the review there was an absurd sentence: "The author claims that the consistency of number theory can be derived from the fundamental conjecture, but this is obviously inconsistent with Gödel's Incompleteness Theorem."

Had Gentzen survived, the history of this area would have been entirely different. This thought makes me feel really regretful. Schütte has said: "Takeuti is the best student of Gentzen." At least I can say that Gentzen was my best teacher. To my regret, I never met him.

Major Figures in Logic

Names	Years
G. Cantor	1845-1918
G.L.G. Frege	1848-1925
D. Hilbert	1862-1943
E.F.F. Zermelo	1871-1953
B. Russell	1872-1970
L.E.J. Brouwer	1881-1966
A.T. Skolem	1887-1963
P. Bernays	1888-1977
A.A. Fraenkel	1891-1965
W. Ackermann	1896-1962
A. Heyting	1898-1980
A. Tarski	1902-1983
A. Church	1903-1995
J. von Neumann	1903-1957
J.H.C. Whitehead	1904-1960
K. Gödel	1906-1978
J. Herbrand	1908-1931
S.C. Kleene	1909-1994
G. Gentzen	1909-1945
A.M. Turing	1912-1954
P. Erdös	1913-1996
A. Robinson	1918-1974
G. Kreisel	1923-
P.J. Cohen	1934-
R.M. Solovay	1938-